Abstract

Calculation of Error Structures in Binary Channels with Memory – the BSC-M Model

Universities teaching Communication Technology are certain to benefit from applications of the BSC-M Model, for the first time facilitating the evaluation of transmission procedures and codes.

This book is written for those interested in the development and testing of link control protocols and any error protection schemes for communication systems. The content is thereby intended to offer a full set of calculations of error structures occurring within binary channels with memory (BSC-M).

As a result it is not necessary to measure error structures. The model contains only two parameters, the bundling factor and the bit error rate, whereby the most diverse range of channels can be presented, derived from empirical readings conducted over twenty-three years. The model has been verified by different authors, as well as by measurements over cable and also over maritime radio up to distances of 7,500 km (short-wave).

www.channels-networks.net
printing and selling by www.bod.de
ISBN 978-3-981-9532-2-0 printed book

Calculations of Error Structures in Binary Channels with Memory

BSC-M as an extension of the BSC Channel

Claus Wilhelm

Third edition 2019

Preface

The book is written for those interested in the development and testing of link control protocols and error protection schemes for communication systems. The content is thereby intended to offer a full set of calculations of error structures occurring within binary channels with memory (BSC-M). The new approach to calculating error structures using the *A-Model* based on its predecessor the *L-Model* is derived from empirical readings conducted over twenty-three years, from 1966-1989, based on a vast number of channels with burst errors, thousands of measuring hours in channels with as few as 200 bit/s up to 2,048 Mbit/s with varying error structure measuring devices. These models have been verified by different authors, as well as by measurements over cable and over maritime radio up to distances of 7,500 km (short-wave).

The current teaching material at universities is dominated by work concerning binary channels without memory whereby the easiest models typically involve:

- Binary Symmetric Channels (BSC),

- Additive White Gaussian Noise (AWGN).

These models do not contain burst structures, as an unrealistic assumption. On the other hand, intuitively derived mathematical models exist with which one can attempt to adapt measured readings, for instance E. N. Gilbert's Model[1] from 1960, commonly termed Hidden Markov Models (HMM) as per William Turin[2]. Therefore by using Markov Models, burst structures can be simulated.

However the distance distribution between bursts cannot be simultaneously simulated, as Models comprise a limited number of states based on finite samples (an example of the fuzzy problem).

Calculations of error structures using these models are rarely found in the relevant literature. One example is the calculation of the probability that g bit errors will be found in a block of length n.

The A-Model is based on a distance function between bit errors, which is similarly found in a large variety of channels. The model contains only **two parameters**, the bundling factor and the bit error rate, whereby the most diverse range of channels can be presented.

Despite the simplifying assumption that the A-Model is a renewal model, by which the memory only reaches the last error symbol, the resulting calculations of probabilities prove to approximate real measured values. With the help of such models, closed

---

[1] Gilbert, E N: "Capacity of a Burst-Noise Channel" in Bell Systems Technical Journal 39 (1960/II), pp 1253-1265.

[2] Turin, William: "Performance Analysis and Modeling of Digital Transmission Systems" 2004, Kluwer Academic / Plenum Publishers.

analytical expressions for error structures of bursts and blocks can be developed. An example is the probability of g errors occurring in bursts or in blocks of length n, or the probability of burst lengths and error patterns.

As a result, the model enables both the comparison of different codes and procedures with one another as well as the optimization of communication systems. It is sufficient to vary the two parameters. For this purpose **it is not necessary** to measure error structures.

Universities teaching Communication Technology are certain to benefit from applications of the A-Model, for the first time facilitating the evaluation of transmission procedures and codes. On introduction is presented in the teaching materials of LNTwww, TU München, from 2016 [1].

For college students, professors and commercial developers finally able to make such evaluations more simply, an entire new realm of possibilities for future problem solving – such as in Machine-to-Machine (M2M) Communication – will come into being. Until now we have been limited to empirical tests, based on already standardized protocols concerning different channels whereby we required no more than a minimum quality for such channels.

Acknowledgements

For their assistance in making the publication of this book possible, I would like to thank Gregor Kendzierski for transforming the MathType formulas into LaTeX and also for designing the book's graphics and tables based on a previous layout designed by Dörthe Reblin, to whom I owe my thanks. Charles Evans helped with the revision of the English version of the book.

I would also like to thank Professor Günter Söder, who has included a summarized version of the book's content as teaching material at the Technical University of Munich as accessible on the University's homepage LNTwww. Professor Söder has incorporated therein a detailed explanation as to how the A-Model extends the binary channel BSC through the bundling factor.

Contents

I. Basics   1

1. A-Model and L-Model for describing BSC-M   3
1.1. Channel disturbance and interruptions 3
1.2. Empirically determined channel properties 4
1.3. Error distance distribution of the L-Model 5
1.4. Generating function and the distribution of error distance in the A-Model   8
1.5. Symbol Error Probability and Block Error Probability 11
1.6. Probability $v(k)$ of the Gap Length k in the A-Model 12
1.7. Symbol correlation function $r(s)$ for the A-Model 12
1.8. Determining Parameters of the A-Model from Measurements 14
1.9. Model Parameter Estimation without Error Structure Measurements . 15
1.10. Summary 17

II. Probabilities of error structures in bursts   19

2. Calculation of error structures in bursts   21
2.1. Number of Bursts 21
2.2. Burst Weight Distribution 22
2.3. Burst Length Distribution 23
2.4. Distribution of Bursts with length l and weight g 29
2.5. Burst Analysis Matrix 31

III. Probability of error structures in blocks   33

3. Calculation of error structures in blocks   35
3.1. Single errors in blocks of length n 36
3.2. Distribution of error length l in errorneus blocks 38
3.3. Distribution of weight g in error bundles of length l 41
3.4. Distribution of weight g of error bundles of length l into erroneous blocks   46
3.5. Distribution of weight g of error bundles in erroneous blocks (weight spectrum) 48

4. Appendix – Basic formulars for the A-Model   51
4.1. Definitions 51
4.2. General relationsships for the A-Model 51

4.3. Burst relationsships 52
4.4. Disturbed block relationsships 52

Bibliography   **53**

List of Figures

1.1. Principle of the error structure measurement 3
1.2. Sequence of the disturbed and interrupted transmission intervals 4
1.3. Model of binary symmetric channels 4
1.4. Parameters for the description of the modeled error sequence 5
1.5. Block error probability as a function of the block length (examples) . . 6
1.6. Calculation of $p_{\mathrm{b}}(n)$ from $u(k)$ 6
1.7. Calculation of the correlation probability $r(s)$ for the occurrence of two symbol errors at a distance of s 12

2.1. Markov-Diagram for calculating the burst weight distribution 22
2.2. Calculation of the burst length distribution; for $r(7)$ 23
2.3. The Burst of length 3 with 3 errors in Example 3 25
2.4. Probability for the burst length l in bursts where $a = 2$ 28

3.1. Calculation of the probability of single errors in a block of length n . . 36
3.2. Error bundle of length l in a block of length n 38
3.3. Calculating the distribution of weight in error bundles of length l . . . 41
3.4. Calculating the probability of the occurrence of error bundles of the length l with g symbol errors in block of length n 46

List of Tables

1.1. Typical values for the parameters of A-Model and L-Model for estimated calculations 9
1.2. Error distance distribution $u(k)$, measured and calculated with the A-Model as the probability that the gap length between two symbol errors is $\geq k$ 15
1.3. Comparison of measured and calculated frequencies of error structures calculated with the A-Model 16

2.1. Burst Analysis Matrix of an artificially generated error sequence ($a = 5$) 31

3.1. Measured and calculated weight distributions in disturbed blocks . . . 35
3.2. Channel matrix of an artificially generated error sequence for bundle errors of length l and weight g in blocks of length $n = 15$ 36
3.3. Length l of error bundles in block of length $g = 5$ 41

Nomenclature

k	Number of error-free symbols between two symbol errors
s	Error distance
$p\{\cdot\}$	Probability
$E\{\cdot\}$	average number of errors in a burst
$E(k)$	average error distance
p_e	Symbol error probability
p_b	Block error probability
$1 - \alpha$	Bundling factor of successive errors
C	Limiting factor for convergence to the BSC-Channel without memory
n	Block length
g	Number of symbol errors in a block, burst or bundle
l	Error burst length in bursts or error bundle length in blocks
a	Number of errorfree symbols to separate two bursts
z_f	Number of symbol errors in the sample
z_b	Number of bursts in the sample
$\Gamma(x)$	Gamma Funktion
$v(k), v_k$	Probability of exactly k error-free steps after a symbol error
$u(k), u_k$	Probability of at least k error-free steps after a symbol error
$r(s)$	Symbol correlation probability
c, A, B	Coefficient
$\mathcal{L}$	Error Bit Sign
$\mathcal{O}$	Error free Bit Sign
$U(t)$	Generation Function of $u(k)$
$R(t)$	Generation Function of $r(s)$
$P(t)$	Generation Function of $p(s)$
$F(t)$	Abbreviated Generation Function of $p(t)$
$V(t)$	Generation Function of $v(k)$
r	Number of Abbreviations
n^*	total signal length

Part I.

Basics

1. A-Model and L-Model for describing BSC-M

Symmetric binary channels with memory can be represented by so-called burst models [2],[3] which satisfy specified accuracy requirements. In this case, the memory length is a function of the memory content. The construction of such models is possible by statistical analysis of measured error sequences [4],[5] with special error structure measuring equipments (Fig. 1.1).

The measurement of errorsequences with the required measuring devices is expensive and time consuming. But for theoretical studies in teaching and research, such models needed for the symmetric binary channel with memory models must be sufficiently precise and mathematically as simple as possible for users. For this reason, we introduce two models, the L-Model[1] (Gap Model) and the A-Model[2] (Distance Model). They form the basis of the clear presentation and calculation of a large number of practically occurring binary channels. In particular, the application of the A-Model allows the calculation of interesting error structures in bursts and blocks using relatively simple formulas. In the following proposed models, the memoryless channel is included as a borderline case. Therefore, it should be dispensed with in most cases, to enable the meaningful use of a model of the binary symmetric channel without memory as the basis of classical considerations.

1.1. Channel disturbance and interruptions

Each received signal which has been transmitted through a disturbed channel of some description, will, at best, be only similar to the original signal, but not identical. The amount of the expected signal forms should therefore be chosen so that the proba-

---

[1]german: Lücken-Modell
[2]german: Abstand-Modell

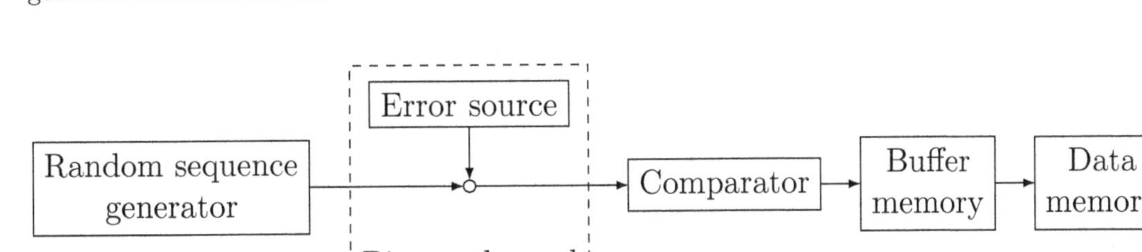

Figure 1.1.: Principle of the error structure measurement

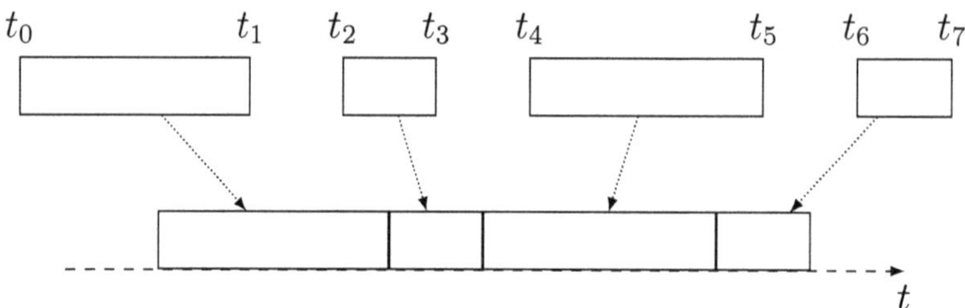

Figure 1.2.: Sequence of the disturbed and interrupted transmission intervals

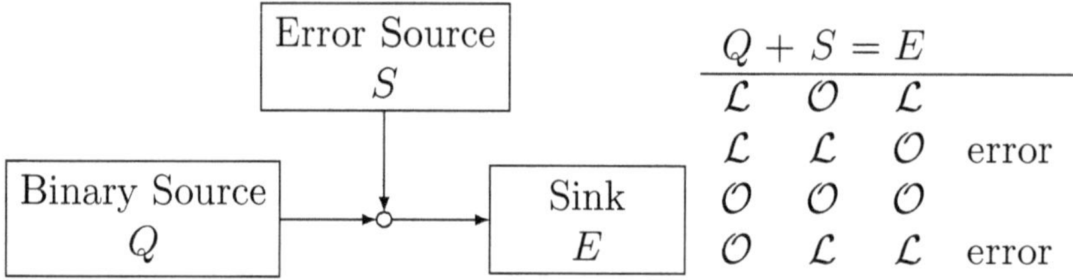

Figure 1.3.: Model of binary symmetric channels

bilities for the random transformation of valid signals into each other be as small as possible. By sampling the amplitude values (symbol recognition) and by comparing them with a reference value, the transmitted binary signals will be recognized as correctly as possible in the receiver. The sampling needs to be located within the symbol. The time interval for the symbol detection through synchronization of the demodulated signals is thereby derived. It is presumed that the channel is interrupted if the synchronization between transmitter and receiver is lost and that during the interruption, the synchronisation cannot be restored. In the following models, interruptions are neglected and intervals are seamlessly joined together (Fig. 1.2).

In the symmetric binary channel, the symbol error generated by the error source S is independent of the symbols of the source Q (Fig. 1.3). The so-defined channel is fully described by the timed order of the occurrence of symbol errors.

1.2. Empirically determined channel properties

Each modeled error sequence contains k errorfree symbols between any two symbol errors. Given the probability $v(k)$, the distance to the next symbol error is exactly k errorfree steps (Fig. 1.4b). The probability $u(k)$ that after a symbol error of at least k errorless steps follows (Fig. 1.4c), is the error distance distribution with

$$u(0) = 1 \text{ and } \lim_{k \to \infty} u(k) = 0$$

$$v(k) = u(k) - u(k+1). \tag{1}$$

If the data are transferred in blocks of length n, these blocks are incorrect with probability $p_\mathrm{b}(n)$. The block error probability is a function of block length n.

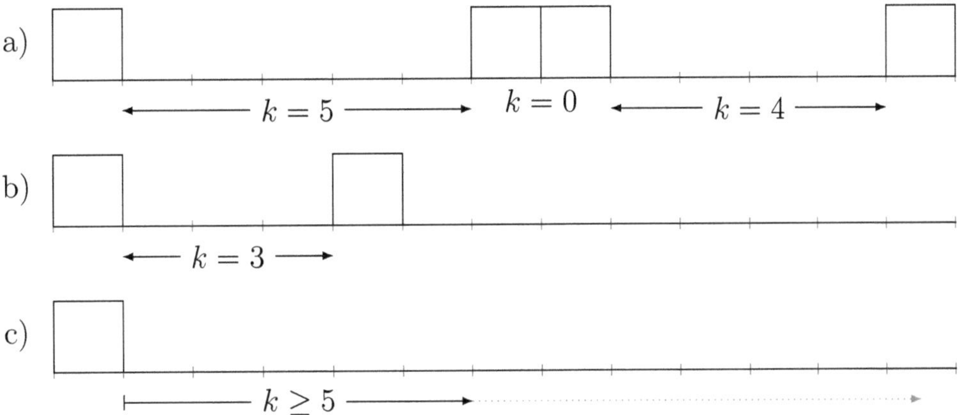

Figure 1.4.: Parameters for the description of the modeled error sequence

Fig. 1.5 shows these functions in doublelogarithmic presentation. In a variety of analyses, the fact that the presented functions show linear increases from about 0.5 to 0.95 in the initial part, was empirically confirmed several times over. This applies for all previously evaluated measurements in the range of 50 bit/s to 2.048 Mb/s. For the asymptotes shown in Fig. 1.5, the linear equation applies

$$\log\left[p_n(n)\right] = \log(p_e) + \alpha \log(n)$$
$$p_{\mathrm{b}}(n) = p_{\mathrm{e}}\, n^\alpha \tag{2}$$

where p_{e} is the symbol error probability, and $(1 - \alpha)$ is the bundling factor, where $0 \le (1 - \alpha) \le 0.5$.

1.3. Error distance distribution of the L-Model

From Fig. 1.6 we can see how the block error probability may be calculated from the error distance distribution. The probability that the first error bit in block of length n is located at some position, is $p_{\mathrm{e}} \cdot u(k)$, where k is the number of error free steps in front of this symbol error. By totaling all n positions we obtain the following result:

$$p_{\mathrm{b}}(n) = p_{\mathrm{e}} \sum_{k=0}^{n-1} u(k) \tag{3}$$

From the empirically found asymptote for p_{b} we then obtain an asymptote for $u(k)$:

$$p_{\mathrm{b}}(n) = p_{\mathrm{e}} \sum_{k=0}^{n-1} u(k) = p_{\mathrm{e}}\, n^\alpha$$

For $n = 1, 2, 3, \ldots$ we obtain by substitution

$$u(0) = 1^\alpha,$$
$$u(0) + u(1) = 2^\alpha \text{ and}$$
$$u(0) + u(1) + u(2) = 3^\alpha.$$

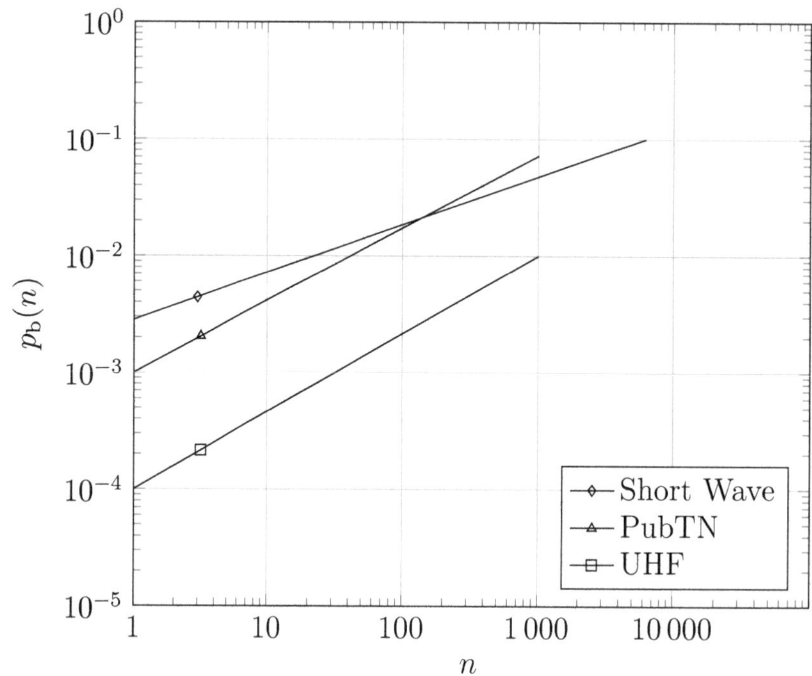

Figure 1.5.: Block error probability as a function of the block length (examples). PubTN: Public Telephone Network; UHF: Ultra-High-Frequency

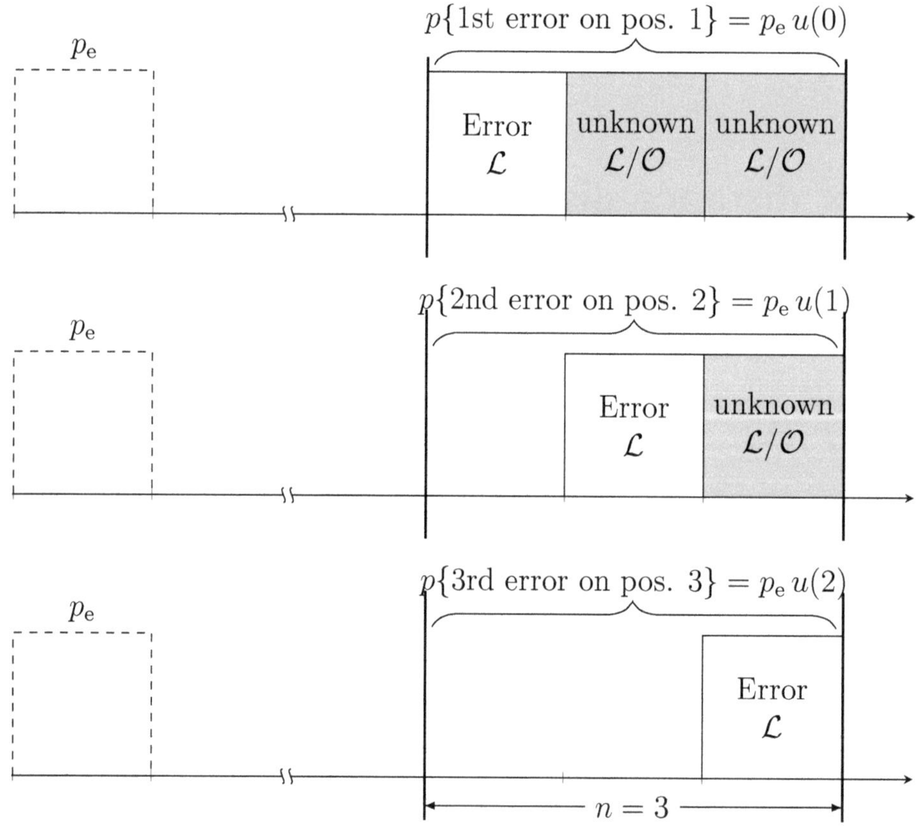

Figure 1.6.: Calculation of $p_\mathrm{b}(n)$ from $u(k)$

It follows then, for the desired asymptote

$$u(k) = (k+1)^\alpha - k^\alpha. \tag{4}$$

However, as the asymptote in accordance with (2) for large block length n values of $p_{\mathrm{b}}(n) > 1$ is achieved, then $u(k)$ is unsuitable as a distribution function according to (4). Therefore it must be

$$\lim_{n \to \infty} p_{\mathrm{e}} \sum_{k=0}^{n-1} u(k) = 1 \quad \text{and} \tag{5}$$

$$\lim_{n \to \infty} p_{\mathrm{b}}(n) = 1$$

To meet the additional condition of (5), $u(k)$ is empirically multiplied by the asymptote $\mathrm{e}^{-\beta k}$

$$u(k) = \left[(k+1)^\alpha - k^\alpha\right] \mathrm{e}^{-\beta k} \tag{6}$$

Here $\beta > 0$. The parameter β must be determined so that equation (5) is satisfied.

Using the approximate generating function, from which the later A-Model has also been developed, we obtain, according to formula (24):

$$C = \mathrm{e}^{-\beta} \equiv \left(1 - p_{\mathrm{e}}^{\frac{1}{\alpha}}\right) \tag{7}$$

Note that the block error probability $p_{\mathrm{b}}(n)$ remains constant when the gaps between the symbol errors are rearranged, so (3) remains unaffected. It follows that for a given distance distribution $u(k)$, the function $p_{\mathrm{b}}(n)$ remains unchanged whether successive gaps occur dependently or independently of one another. Therefore, it is assumed for the sake of simplicity, that the gaps are independent.

Definition 1: The L-Model

The L-Model (Gap Model) of the binary symmetric channel with memory contains successively independent gaps of length k between the individual error symbols. The probability $u(k)$ that after a symbol error $\geq k$ error-free symbols follow, is calculated from

$$u(k) = \left[(k+1)^\alpha - k^\alpha\right] C^k \quad \text{where } C = \mathrm{e}^{-\beta} \equiv (1 - p_{\mathrm{e}}^{(1/\alpha)}), \text{ from fromula (24)} \tag{8}$$

where $u(0) = 1$; $\lim_{k \to \infty} u(k) = 0$; k is the number of error free steps between two symbol-errors, $(1 - \alpha)$ is the bundling factor in the channel $0.5 \leq \alpha \leq 1$.

In this model, the channel-memory extends back to the previous symbol error in each case. The gap lengths k are independently and identically distributed random variables. The symbol errors are **recurrent events**, whereby Markov's chains occurring within bursts are dispensed with.

1.4. Generating function and the distribution of error distance in the A-Model

The L-Model is used to calculate the probability of the occurrence of error patterns within a time interval Δt. Each error pattern is one of 2^b elementary events which can be observed within a time segment of bits-length b.

However, for practical use of channel models, it is necessary to calculate the probabilities for the occurrence of error patterns. Error structures are subsets of error patterns which have the same predetermined characteristics, such as the number g of symbol errors, the length l of the burst error, and so on. To calculate the probabilities of such sums of random variables, the so-called generating function should be employed. In this specific case, the gap length k is our random variable; $u(k)$ is designated as u_k, the probability between any two symbol errors, when the gap is equal to or more then k bits. Therefore

$$U(t) = u_0 + u_1 t + u_2 t^2 + \cdots = \sum_{k=0}^{\infty} u_k t^k \tag{9}$$

is the generating function of the error distance distribution $u(k)$. If the integers k_1 and k_2, occur as mutually exclusive and independent, as well as non-negative random variables with generating functions $U_1(t)$ and $U_2(t)$, then sum $k_1 + k_2$ is the generating function:

$$C(t) = \sum_{k=0}^{\infty} C_k t^k = U_1(t) U_2(t) \tag{10}$$

In the following, a generating function is derived as an approximation for the previously derived L-Model, whose regression provides, for the sake of analysis, an excellent application of our so-called A-Model. Our A-Model approaches the real binary channel as does our L-Model. The generating function of the L-Model, which results from (6), is derived as:

$$U(t) = \sum_{k=0}^{\infty} [(k + 1)^\alpha - k^\alpha] \, \mathrm{e}^{-\beta k} \, t^k. \tag{11}$$

At the present time, a closed analytical expression for the sum represented by equation (11) cannot be specified. Consequently, the following approximated solution is derived for (11).

$$(k + \Delta k)^\alpha = k^\alpha \left(1 + \frac{\alpha}{1!k} \Delta k + \frac{\alpha}{1!k^2} \Delta k^2 + \dots \right) \tag{12}$$

Using the first two summands above, where $\Delta k = 1$, the following is obtained:

$$(k + \Delta k)^\alpha - k^\alpha \approx k^\alpha \frac{\alpha}{k} = \alpha \, k^{(\alpha-1)} \tag{13}$$

Further, it is approximated in (11) that

$$U(t, k) = c \int_{k=0}^{\infty} \alpha \, k^{(\alpha-1)} \, \mathrm{e}^{-\beta k} \, t^k \mathrm{d}k$$

$$= c\,\alpha \int_{k=0}^{\infty} k^{(\alpha-1)} \, \mathrm{e}^{k(\ln t - \beta)} \mathrm{d}k$$

Table 1.1.: Typical values for the parameters of A-Model and L-Model for estimated calculations

Modelling Data transmission via	Error probability	Bundling factor
Short Wave	$5 \cdot 10^{-2} > p_e > 5 \cdot 10^{-3}$	$0.5 > (1 - \alpha) > 0.25$
APC (med. speed; mech. selectors)	$1 \cdot 10^{-2} > p_e > 1 \cdot 10^{-4}$	$0.3 > (1 - \alpha) > 0.1$
APC (200 Bd; mech. selectors)	$1 \cdot 10^{-3} > p_e > 1 \cdot 10^{-5}$	$0.3 > (1 - \alpha) > 0.1$
Leased Line (48 kbit/s with modem)	$1 \cdot 10^{-5} > p_e > 1 \cdot 10^{-6}$	$0.3 > (1 - \alpha) > 0.1$
PCM-C (with 2,048 Mbit/s)	$1 \cdot 10^{-7} > p_e > 1 \cdot 10^{-9}$	$0.2 > (1 - \alpha) > 0$

APC: Automatic Phone Channels; PCM-C: Puls Code Modulation-Channels

The constant c should be chosen later to be as consistent as possible, by finding in [6]:

$$\int_{k=0}^{\infty} e^{-\omega k} k^{(\delta-1)} \mathrm{d}k = \frac{\Gamma(\delta)}{\omega^{\delta}} \quad \text{where } \mathrm{Re}(\omega) > 0, \mathrm{Re}(\delta) > 0 \tag{14}$$

It follows:

$$U(t) = c\,\alpha \frac{\Gamma(\alpha)}{(\beta - \ln t)^{\alpha}}. \tag{15}$$

In the closed environment of $t = 1$, concerning normal technical channels of interest in Tab. 1.1: $0 < p_e < 10^{-2}$; $0 < \beta \le p_e$, the following applies:

$$|\beta - \ln t| \ll 1$$
$$\beta - \ln t = 1 - (1 - \beta + \ln t)$$
$$\beta - \ln t \approx 1 - e^{-\beta} t.$$

Thereby we obtain:

$$U(t) = c\,\alpha\,\Gamma(\alpha) \frac{1}{(1 - e^{-\beta} t)^{\alpha}} \quad \text{where } 1 - \Delta < t < e^{\beta}; \Delta \ll 1; \alpha > 0; \beta \ll 1. \tag{16}$$

Since (16) can be differentiated at any point of the interval $t < e^{\beta}$ as often as desired, $U(t)$ can be developed at the position $t_0 = 0$ in a power series, which converges in the close environment of $t = 1$:

$$U(t) = U(0)\,t^0 + \frac{U'(0)}{1!}\,t^1 + \frac{U''(0)}{2!}\,t^2 + \cdots \tag{17}$$

or

$$U(t) = u(0)\,t^0 + u(1)\,t^1 + u(2)\,t^2 + \cdots.$$

According to (1), $u(0) = 1$. It follows

$$U(0) = 1 = c\,\alpha\,\Gamma(\alpha) \frac{1}{(1 - e^{-\beta} 0)^{\alpha}}.$$

Therefore, our desired generating function is

$$U(t) = \sum_{k=0}^{\infty} u(k)\,t^k = \frac{1}{(1 - C t)^{\alpha}} \quad \text{for } 1 - \Delta < t < e^{\beta}; \Delta \ll 1; \alpha > 0; \beta \ll 1. \tag{18}$$

The Taylor Series Expansion as in (17), leads to the new approximated error distance distribution based on the L-Model (6)

$$u(k) = \frac{\alpha(1+\alpha)\cdots(k-1+\alpha)}{k!}\mathrm{e}^{-\beta k} = \prod_{i=0}^{k-1}\frac{i+\alpha}{k!}\mathrm{e}^{-\beta k} \quad \text{for } k \geq 1 \text{ with } u(0) = 1.$$

(19)

As given in [6], the following equations apply to the gamma function Γ.

$$\alpha(1+\alpha)\cdots(k-1+\alpha) = \prod_{i=0}^{k-1}(i+\alpha) = \frac{\Gamma(\alpha+k)}{\Gamma(\alpha)} \quad \text{with } k! = \Gamma(k+1)$$

$$u(k) = \frac{\Gamma(\alpha+k)}{\Gamma(1+k)\,\Gamma(\alpha)}\mathrm{e}^{-\beta k} \quad \text{with } \mathrm{e}^{-\beta} \equiv (1-p_\mathrm{e}^{\frac{1}{\alpha}}) = C, \qquad (19\mathrm{a})$$

derived from fromular (24).

> ### Definition 2: The A-Model
>
> A channel model with independently successive gaps between symbol errors which has the distribution function for the error distances
>
> $$u(k) = \frac{\alpha(1+\alpha)\ldots(k-1+\alpha)}{k!}\,C^k = \frac{C^k}{k!}\prod_{i=0}^{k-1}(i+\alpha) \qquad (20)$$
>
> whereby $C = (1-p_\mathrm{e}^{\frac{1}{\alpha}})$ according to formula (24). Here, $k = 1, 2, \ldots$ is the distance between two symbol errors, and $u(0) = 1$, and p_e the symbol error probability and $(1-\alpha)$ the bundling factor. According to the formula (24), the A-Model includes the memory-free BSC channel on account of $u(k; \alpha = 1) = (1 - p_\mathrm{e})^k$. Recursively $u(k)$ can be calculated thus:
>
> $$u(k+1) = u(k)\,\frac{k+\alpha}{k+1}\,C \quad \text{with } u(0) = 1.$$
>
> And the genration function of $u(k)$ is:
>
> $$U(t) = \sum_{k=0}^{\infty} u(k)\,t^k = \frac{1}{(1-C\,t)^\alpha}.$$

Approximation for $u(k)$ in the A-Model for large values of k:

Here an approximation for the quotient $\frac{\Gamma(\alpha+k)}{\Gamma(1+k)}$ for large values of k is needed with the estimate [6]

$$y(k) = \ln\Gamma(k) \approx (k - 0.5)\ln(k) - k + 0.5\ln(2\pi) \tag{21}$$

$$y'(k) = \ln k + \frac{k - 0.5}{k} - 1 = \ln k - \frac{1}{2k}$$

$$y(k + \Delta k) \approx \ln\Gamma(k) + \left(\ln k - \frac{1}{2k}\right)\Delta k$$

$$\frac{\Gamma(\alpha + k)}{\Gamma(1 + k)} \approx \frac{\exp\left[\ln\Gamma(k)\right]}{\exp\left[\ln\Gamma(k)\right]} \frac{\exp\left[\alpha\left(\ln k - \frac{1}{2k}\right)\right]}{\exp\left[1\left(\ln k - \frac{1}{2k}\right)\right]} = \exp\left[-(1 - \alpha)\ln k + \frac{1 - \alpha}{2k}\right]$$

$$u(k) \approx \frac{1}{\Gamma(\alpha)} k^{\alpha-1} C^k e^{\frac{1-\alpha}{2k}} \approx \frac{1}{\Gamma(\alpha)} k^{\alpha-1} C^k \quad \text{for } k > 10; k \ll 10 \tag{22}$$

$\Gamma(\alpha)$ is set with the help of `Excel` as per "$=Exp(GammaLn(\alpha))$". Example for $\alpha = 0.7$; $p_e = 0.01$ is $u(1000) = 0.024149$ recursively and 0.024143 approximately.

1.5. Symbol Error Probability and Block Error Probability

The reciprocal of the symbol error probability p_e is calculated from the average error distance $E(k)$:

$$\frac{1}{p_e} = E(k) + 1$$

$$E(k, t = 1) = \sum_{k=0}^{\infty} k\, v(k)\, t^k = \left(\sum_{k=0}^{\infty} u(k)\, 1^k\right) - 1 = U(1) - 1 \tag{23}$$

$$p_e = \frac{1}{U(1)} = \left(1 - e^{-\beta}\right)^{\alpha} \quad \text{whereby } p_e^{\frac{1}{\alpha}} = 1 - e^{-\beta} \text{ and } C = e^{-\beta} \equiv \left(1 - p_e^{\frac{1}{\alpha}}\right) \tag{24}$$

The block error probability is, see (8) and (20):

$$p_b(n) = p_e \sum_{k=0}^{n-1} u(k)$$

For the BSC Channel (without memory), where $\alpha = 1$ results for the L-Model and the A-Model:

$$p_b(n, \alpha = 1) = p_e \left(u(0) + \sum_{k=1}^{n-1} u(k)\right) = p_e \left(1 + \sum_{k=1}^{n-1} C^k\right) = p_e \frac{1 - (1 - p_e)^n}{1 - (1 - p_e)}$$

$$= 1 - (1 - p_e)^n$$

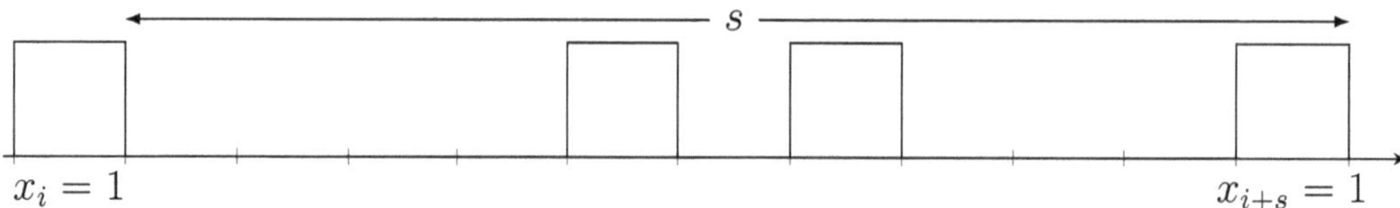

Figure 1.7.: Calculation of the correlation probability $r(s)$ for the occurrence of two symbol errors at a distance of s

1.6. Probability $v(k)$ of the Gap Length k in the A-Model

Because $v(k) = u(k) - u(k+1)$, it follows from (19)

$$v(k) = u(k) \left(1 - \frac{k+\alpha}{k+1} e^{-\beta} \right) \tag{25}$$

$$v(k) = u(k) \left(1 - \frac{k+\alpha}{k+1} C \right). \tag{26}$$

For $v(k; \alpha = 1) = u(k)\, p_\mathrm{e} = p_\mathrm{e} \,(1 - p_\mathrm{e})^k$ the the memoryless BSC Channel with $\alpha = 1$ results.

1.7. Symbol correlation function $r(s)$ for the A-Model

In various cases of practical interest, the probability of error structures should be calculated whereby one symbol error occurs exactly s steps after another symbol error. Whether or not the intervening symbols are also wrong is not relevant Fig. 1.7. An example is the calculation of the probability of the length of error bundles in erroneous blocks. For these applications, the correlation probability $r(s)$ is determined for the A-Model when two symbol errors occur at a distance s at the points i and $i + s$ of the error sequence: $r(s) = p\left\{x_i = \mathcal{L}; \ x_{i+s} = \mathcal{L}\right\}$

For our calculation, we denote the probability $p(s) = v(s-1)$ for the first return of an error after exactly s steps at a distance of s with $p(s) = p_s$. For the n-th return of a symbol error we write $p_s^{(n)}$ where $p_s^{(n)} = 0$ for $n > s$. This gives us:

$$r(s) = p_\mathrm{e} \left(p_s^{(1)} + p_s^{(2)} + \ldots \right). \tag{27}$$

Multiplied by t^s and summed over all s, we obtain the generating function for the correlation probability:

$$R(t) = \sum_{s=1}^{\infty} r(s)\, t^s = p_\mathrm{e} \left(\sum_{s=1}^{\infty} p_s^{(1)}\, t^s + \sum_{s=1}^{\infty} p_s^{(2)}\, t^s + \ldots \right). \tag{28}$$

Now the correlation probability for 0 steps is added:

$$r(0) = p_\mathrm{e} \cdot 1$$

$$p_0^{(1)} = p_0^{(2)} = \cdots = 0$$

And we apply the convolution theorem:

$$\sum_{s=0}^{\infty} p_s^{(n)} \, t^s = [P(t)]^n \tag{29}$$

Here is $P(t)$ the generating function for the probability of the first recurrence of a symbol error after s steps. The result is the generating function $R(t)$ for the correlation probability $r(s)$:

$$R(t) = \frac{p_{\mathrm{e}}}{1 - t \, V(t)},$$

$$R(t) = \sum_{s=0}^{\infty} r(s) \, t^s = p_{\mathrm{e}} \left(1 + P(t) + [P(t)]^2 + [P(t)]^3 + \ldots \right). \tag{30}$$

The geometric series is added, and we obtain:

$$R(t) = \frac{p_{\mathrm{e}}}{1 - P(t)}.$$

According to Fig. 1.7,

$$p(s) = v(s - 1).$$

This results in $p(0) = 0$ from the formula

$$P(t) = \sum_{s=1}^{\infty} p(s) \, t^s = \sum_{s=1}^{\infty} v(s - 1) \, t^s, \text{ and } P(t) = \sum_{k=0}^{\infty} v(k) \, t^{k+1} = t \sum_{k=0}^{\infty} v(k) \, t^k.$$

Remembering the simple context $P(t) = t \, V(t)$ then

$$R(t) = \frac{p_{\mathrm{e}}}{1 - t \, V(t)} \tag{31}$$

is the generating function for the correlation probability $r(s)$. From [3], (3.61) and (3.97) we obtain

$$V(t) = \frac{1}{t} + \frac{t - 1}{t \, (1 - C \, t)^{\alpha}}, \quad 1 - t \, V(t) = \frac{1 - t}{(1 - C \, t)^{\alpha}} \tag{31a}$$

From this, the generating function $R(t)$ for the A-Model becames:

$$R(t) = p_{\mathrm{e}} \frac{1 - C \, t}{1 - t} \tag{32}$$

For the determination of $r(s)$, $R(t)$ should be developed as a power series and ordered according to powers of t_s. The series expansion of the counter at the position $t = 0$ yields in steps:

$$\begin{aligned}
y &= (1 - C \, t); & y(0) &= 1; \\
y' &= -\alpha(1 - C \, t)^{\alpha - 1} \, C; & y'(0) &= -\alpha \, C; \\
y'' &= -\alpha(1 - \alpha)(1 - C \, t)^{\alpha - 2} \, C^2; & y''(0) &= -\alpha(1 - \alpha) \, C^2; \\
y''' &= -\alpha(1 - \alpha)(2 - \alpha)(1 - C \, t)^{\alpha - 3} \, C^3; & y'''(0) &= -\alpha(1 - \alpha)(2 - \alpha) \, C^3;
\end{aligned} \tag{33}$$

$$y = 1 - \frac{\alpha\,C}{1!}\,t - \frac{\alpha(1-\alpha)}{2!}\,C^2 t^2 - \frac{\alpha(1-\alpha)(2-\alpha)}{3!}\,C^3 t^3 - \dots$$

For the denominator we can write:

$$\frac{1}{1-t} = 1 + t + t^2 + t^3 + \dots \tag{34}$$

With the approach

$$R(t) = \sum_{s=0}^{\infty} r(s)\,t^s = r(0)\,t^0 + r(1)\,t^1 + r(2)\,t^2 + \dots$$

and by the multiplication of the two series (33) and (34), ordering of powers of t_s and comparison of the coefficients, the desired solutions are directly obtained:

$$r(0) = p_{\mathrm{e}} \cdot 1,$$
$$r(1) = p_{\mathrm{e}}\left(1 - \frac{\alpha}{1!}\,C\right),$$
$$r(2) = p_{\mathrm{e}}\left(1 - \frac{\alpha}{1!}\,C - \frac{\alpha(1-\alpha)}{2!}\,C^2\right) \text{ and}$$
$$r(s) = p_{\mathrm{e}}\left(1 - \frac{\alpha}{1!}\,C - \frac{\alpha(1-\alpha)}{2!}\,C^2 - \dots - \frac{\alpha(1-\alpha)\dots(s-1-\alpha)}{s!}\,C^s\right) \tag{35}$$

as the correlation probability for the A-Model. For the symmetrical binary channel without memory, the result with $\alpha = 1$ is as expected:

$$r(s) = p_{\mathrm{e}}(1 - C) = p_{\mathrm{e}}\left(1 - \left(1 - p_{\mathrm{e}}^{\frac{1}{\alpha}}\right)\right) = p_{\mathrm{e}}^2$$

1.8. Determining Parameters of the A-Model from Measurements

To assess the accuracy of approximation of the A-Model, a measured error sequence was used [3]. It was registered during on a Measurement time of 16 hours. The sample was transmitted from Dresden-Neustadt to Berlin with 200 Baud over the public telephone network (rotary selectors). There were 32 samples strung together, each 30 minutes long, while at the same receiving level $-34.5\,\mathrm{dB}$ of 11,554,541 transmitted symbols in the total sample, 9807 were erroneous ($p_{\mathrm{e}} = 0.0008488$). There were 1560 gaps of length $k = 0$ and 3513 gaps of length $k \geq 100$, i.e. $u(100) = 0.3582$. With the assumed measured value p_{e} for the A-Model, only one unknown parameter for the determination of $u(k)$ with (20) is the bundling factor $(1-\alpha)$ remains. This parameter can be determined using the best estimation method. Since $u(1) = \alpha\left(1 - p_{\mathrm{e}}^{\frac{1}{\alpha}}\right) = 8247/9807 = 0.8409$ the first estimate is $\alpha_0 = 0.84$. On the other hand, $u(k)$ should be

Table 1.2.: Error distance distribution $u(k)$, measured and calculated with the A-Model as the probability that the gap length between two symbol errors is $\geq k$

k	$u(k)$ [7] measured	$u(k)$ [†] calculated
1	0.8409	0.8099
5	0.6393	0.6287
100	0.3582	0.3582
500	0.2492	0.2457
1000	0.2073	0.1986

[†] based on (20), (22)

exactly the value $u(100) = 0.3582$. For this purpose, with the help of `Excel`, a table up to $u(100)$ iteratively calculated

$$u(k+1) = u(k)\frac{k+\alpha}{k+1}\,C \quad \text{where the initial value is } u(0) = 1. \tag{36}$$

We set this values as $p_e = 0.0008488$ and $\alpha_0 = 0.8409$. This given the initial value $u_0(100) = 0.4189142593$. We are looking α for $u(100) = 0.3582$. Using the `Excel` function "Extras/Goal Seek" (multiple times) and gradually looking for the findings obtained, yields $\alpha = 0.811511995 \rightarrow \alpha = 0.81145$ and $u(1) = 0.8113$.

The model is therefore accurate for $u(100)$. But it deviates somewhat for small values of k, where $u(1) = 0.8401$ was measured; in this model however $u(1) = 0.8113$. This latter value is used for the following examples. However in Wilhelm [3] 0.84 is used.

1.9. Model Parameter Estimation without Error Structure Measurements

To calculate the effectiveness of data transmission procedures and coding for error protection, channel model must be available, in which for simulation purposes artificial error sequences can be generated, or error structure probabilities can be calculated. For the production of artificial error sequences, the A-Model and L-Model are equally suitable. For the calculation of probabilities of error sequences in particular, the A-Model is suitable because its generating function for the error distance distribution according to (20) is known.

The results of error structure measurements are not required if the two parameters, symbol error probability p_e and bundling-factor $(1 - \alpha)$ are suitably varied. (Tab. 1.3)

Table 1.3.: Comparison of measured and calculated frequencies of error structures calculated with the A-Model

	Formula	Parameter	measured[†]	calculated[‡]
Error structures in Bursts:				
Number z_b of bursts in sample	(37)	$a = 1$	8247	8238
		$a = 2$		7577
		$a = 5$	6270	6659
Mean number of errors $E\{g\}$ in bursts	(38)	$a = 5$	1.56	1.47
Number of bursts with g errors	(40)	$a = 5; g = 4$ not analysed		149
Number of bursts of length l	(57)	$a = 2; l = 5$ not analysed		61
Error structures in Blocks:				
Number of blocks with single errors $(g = 1)$	(68)	$n = 5$	1096	940
		$n = 1023$	6271	6343
Number of blocks with error of length $l \geq 3$	(72)	$n = 5; l \geq 3$	426	319
Number of blocks with error bundle of length $l = n$ and g errors	(88),(89)	$l = n = 5,$ $g = 3$	45	35
Number of blocks of length n with error bundle of length l and g errors	(92)	$l = 5; g = 3;$ $n = 63$	18	48

Burst starts with a symbol error and is completed as soon as a error-free consecutive steps follow.

Bundle in a block starts with a first symbol error and ends with a last symbol error.

[†] sample from a public Strowger switch telephone network with failure-prone mechanical switches, modem 200 Bd, 16 Hours, 9807 symbol errors $p_\mathrm{e} = 0.0008488$

[‡] with A-Model ($\alpha = 0.84$) and same sample properties

1.10. Summary

Using the two presented channel models with independent gaps between successive symbol errors, theoretical assessments of the effectiveness of data transfer procedures and encodings are possible.

According to the given distribution function one can create artificial error sequences and sort, or calculate analytically, their probability for error structures. With both these models, the course of the block probability is reflected exactly asymptotically along the block length. Both models are significantly more accurate than the previously proposed models with independent gaps between consecutive errors. They are even more exact than the well-known principal models of other authors that have been in use for a long time, including Gilbert (1960, [8]).

In conclusion, error structures are presented in Tab. 1.3 for comparison between measured and calculated frequencies of A-Model error structures.

Part II.

Probabilities of error structures in bursts

2. Calculation of error structures in bursts

A burst begins with a symbol error and ends when at least a error free steps follow each other (see Fig. 2.1). Here a is referred to as the distance parameter. The error structures within bursts and the number of bursts are especially of interest with respect to block-free codes [3], [9].

2.1. Number of Bursts

Hereafter the number z_b of bursts is calculated within a sample with z_f symbol errors and the distance parameter a. When the gap for a symbol error is equal to or greater than a, the burst is terminated.

The proportion of these gaps is $u(a)$. The results, regardless of the memory properties of the channel, for each sample yield:

$$z_\mathrm{b} = z_\mathrm{f}\, u(a)$$

The number of bursts remains constant, for all of the permutations of the successive error intervals and for all the error-distance distributions $u(k)$, for which $u(a)$ has the same value. Especially with respect to (20) in part I, the following applies for our A-Model

$$z_\mathrm{b} = z_\mathrm{f}\, \frac{\alpha(1+\alpha)\ldots(a-1+\alpha)}{a!}\, C^a. \tag{37}$$

It is evident that the average number $E\{g\}$ of errors in a burst is therefore easy to calculate:

$$E\{g\} = \frac{z_\mathrm{f}}{z_\mathrm{b}} = \frac{1}{u(a)}. \tag{38}$$

Example 1

Calculate the number z_b of bursts for the A-Model, for a given sample where $z_\mathrm{f} = 9807$, $\alpha = 0.84$ and $p_\mathrm{e} = 0.0008488$, when the distanceparameter is $a = 1$ or $a = 5$.

Solution: Comparatively 8247 or 6270 bursts were measured by sorting the registered sample. In bursts with the parameter $a = 5$, we found an average of 1.47

symbol errors.

$$a = 1: \quad z_{\mathrm{b}} = z_{\mathrm{f}}\,\alpha\,C \approx 9807 \cdot 0.84 = 8238,$$

$$a = 5: \quad z_{\mathrm{b}} = z_{\mathrm{f}}\,\frac{\alpha(1 + 0.84)(2 + 0.84)(3 + 0.84)(4 + 0.84)}{5!}\,C^{-5}$$

$$z_{\mathrm{b}} = 9807 \cdot 0.6798\,\frac{1}{1.0011} = 6659.$$

2.2. Burst Weight Distribution

When we want to calculate how the number of errors is distributed within bursts, we start from Fig. 2.1.

$$1 - u(a) \quad B_i \quad \xrightarrow{u(a)} \quad B_{i+1}$$

Figure 2.1.: Markov-Diagram for calculating the burst weight distribution

The channel is in the state B_i as long as the errors which occur are part of the burst. The Markov chain starts from the state B_i; that is, it is assumed that the burst has already begun with the first error. The errors produced fall into burst B_i as long as the gaps are shorter than a error-free steps. After the "jump" to the next error, the burst can be terminated when the gap is $k \geq a$. The channel then proceeds in the state B_{i+1}, i.e. the next burst. It easily follows then that the weight distribution in bursts for the assumption regarding the independence of gaps is:

$$p\{g = 1\} = u(a),$$
$$p\{g = 2\} = u(a)\,[1 - u(a)],$$
$$p\{g\} = u(a)\,[1 - u(a)]^{g-1} \tag{39}$$

Example 2

Caculate the probability that a burst contains 4 symbol errors, when the distance parameter a has the value 5. Use the model as given in Example 1.

Solution: According to Example 1, the value of 0.6791 was calculated for $u(5)$. Given (39) we obtain

$$p\{g = 4\} = 0.6791\,[1 - 0.6791]^{4-1} = 0.02244 \tag{40}$$

Only 2.244% of the bursts have 4 errors. Means 149 bursts from a total of 6659 (compare with Example 1).

2.3. **Burst Length Distribution**

Under the assumption of independent gaps between symbol errors, the probabilities should be calculated for the burst length l (Fig. 2.2). The correlation $r(7)$ between the first and last errors of the burst is demonstrated in Fig. 1.7, where l is the burst length and a is the distance parameter. First, the correlation is determined under the condition that the gaps between the error positions are always smaller than a.

Using the general approach for the generating function $R(t)$, in determining the correlation between the first and the last errors,

$$R(t) = \frac{p_e}{1 - P(t)} = p_e(1 - C\,t)^\alpha 1 - t \tag{41}$$

It is necessary to determine the new generating function for the first recurrence of a symbol error after s steps within the burst. If the probability of recurrence of a symbol error after s steps within the burst, then $p(s) = 0$ for $s = a$. It follows that the generating function reduces to

$$F(t) = \sum_{s=1}^{a} p_s\, t^s = \sum_{k=0}^{a-1} v(k)\, t^{k+1} \tag{42}$$

This is the so-called "abbreviated generating function" since

$$p(s) = v(s-1) = p_s. \tag{43}$$

Therefore, the generating function for the correlation between the first and the last errors of a burst is

$$R(t) = \frac{p_e}{1 - \sum_{k=0}^{a-1} v(k)\, t^{k+1}}, \tag{44}$$

$$R(t) = \frac{p_e}{1 - v_0\,t - v_1\,t^2 - \cdots - v_{a-1}\,t^a}. \tag{45}$$

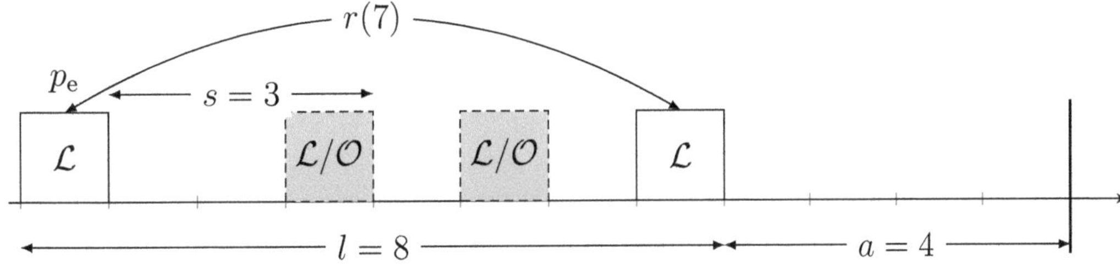

Figure 2.2.: Calculation of the burst length distribution; for $r(7)$

The occurrence of the first error symbol sees the burst already started. The burst is only completed when after the last error symbol at least a error-free symbols follow. This gives the generating function

$$B_a(t) = \frac{u(a)}{1 - \sum_{k=0}^{a-1} v(k)\, t^{k+1}} \tag{46}$$

for the probabilities $p\{l, a\}$ were the burst length is l. The desired probabilities are then the coefficients of t^{l-1} in the power series expansion of $B_a(t)$. The following then applies

$$\sum_{l=1}^{\infty} p\{l, a\} = 1. \tag{47}$$

The corresponding values are to be used for $u(a)$ and $v(k)$ in the A-Model. Hereafter it will be shown that the calculations become more complicated as a increases in value. The burst should finish when an error-free symbol occurs. Then follows

$$B_1(t) = \frac{u(1)}{1 - v_0\, t}. \tag{48}$$

From the expression

$$\frac{1}{1 - x} = 1 + x + x^2 + x^3 + \dots,$$

the power series expansion $B_1(t)$ is generated as follows:

$$B_1(t) = u(1) \left[1 \cdot t^0 + v_0\, t^1 + v_0^2\, t^2 + \dots \right] \tag{49}$$
$$p\{l = 3, a = 1\} = u(1)\, v_0^2.$$

Example 3

Determine the probability that a burst has the length $l = 3$ whereby $a = 1$.

Solution: The required probability is the coefficient of $t^{l-1} = t^2$ in the expression (49)

$$p\{l = 3, a = 1\} = u(1)\, v_0^2$$

For this example, the solution is obtained intuitively (see Fig. 2.3).

Since the weight is also equal to 3, the same solution as in (39) must apply, considering that

$$1 - u(1) = v(0)$$

For the model (Example 1), it is given

$$u(1) = \alpha\, C = 0.81113; \quad v(0) = 0.18868;$$

$$p\{l = 3, a = 1\} = 0.02888$$

From 7958 bursts (see Example 1), 230 have the weight 3.

A burst ends when **two error-free** symbols follow one another. It follows from (46),

$$B_2(t) = \frac{u(2)}{1 - v_0\,t - v_1\,t^2} \tag{50}$$

The power series expansion (obtained by simple division)

$$1/\left(1 - v_0\,t - v_1\,t^2\right) = 1 + v_0\,t + \left(v_1 + v_0^2\right)t^2 + \left(2v_1v_0 + v_0^3\right)t^3 + \dots$$

$$\underline{1 - v_0\,t - v_1\,t^2}$$
$$v_0\,t + v_1\,t^2$$
$$\underline{v_0\,t - v_0^2 t^2 - v_1v_0\,t^3}$$
$$\left(v1 + v_0^2\right)t^2 + v_1v_0\,t^3$$
$$\underline{\left(v_1 + v_0^2\right)t^2 - \left(v_1v_0 + v_0^3\right)t^3 - \dots}$$
$$\left(2v_1v_0 + v_0^3\right)t^3$$

shows that this elementary calculation method contains more complicated coefficients for larger burst lengths l. Therefore, the coefficients are better determined by partial fraction decomposition and the residual method, or they should be estimated asymptotically. For the decomposition into partial fractions, we re-write:

$$B_2(t) = \frac{u(2)}{v_1} \cdot \frac{v_1}{1 - v_0\,t - v_1\,t^2} = \frac{A}{t - t_1} + \frac{B}{t - t_2}. \tag{51}$$

The x-intercept values of the denominator of (51) arise from the following approach:

$$t^2 - \frac{v_0}{v_1}\,t - \frac{1}{v_1} = 0,$$

$$t_{1/2} = -\frac{v_0}{2v_1} \pm \sqrt{\frac{v_0^2}{4v_1^2} + \frac{1}{v_1}} \tag{52}$$

For the determination of the constants A and B, a comparison of the coefficients is carried out.

$$\frac{u(2)}{v_1} = A(t - t_2) + B(t - t_1)$$

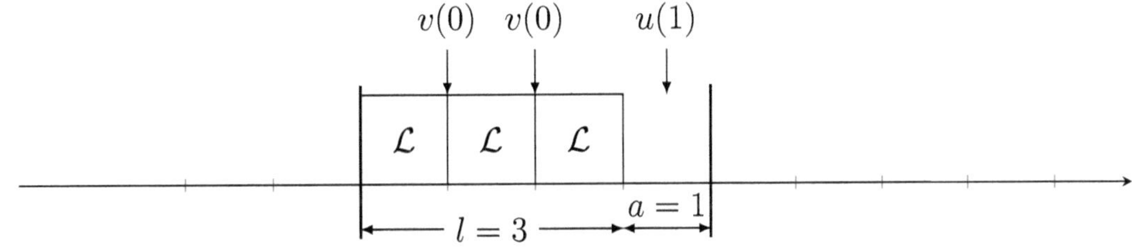

Figure 2.3.: The Burst of length 3 with 3 errors in Example 3

$$A = -B$$

$$-t_2 A + t_1 A = \frac{u(2)}{v_1}$$

$$A = \frac{u(2)}{v_1} \frac{1}{t_1 - t_2}; \quad B = \frac{u(2)}{v_1} \frac{1}{t_2 - t_1} \tag{53}$$

$$t_1 - t_2 = \frac{-u(2)}{\sqrt{v_0^2 + 4v_1}} \tag{54}$$

$$A = \frac{u(2)}{\sqrt{v_0^2 + 4v_1}}; \quad B = -\frac{u(2)}{\sqrt{v_0^2 + 4v_1}} \tag{55}$$

From the power series development (51), the coefficient b_s results by determining the residuals:

$$b_s = -\operatorname*{Res}_{t=t_1} \left\{ \frac{A}{t^{s+1}(t - t_1)} \right\} - \operatorname*{Res}_{t=t_2} \left\{ \frac{B}{t^{s+1}(t - t_2)} \right\}$$

$$b_s = \frac{A}{t_1^{s+1}} + \frac{B}{t_2^{s+1}}. \tag{56}$$

From (55), given $l = s + 1$, the probability that a commenced burst has the length l, is

$$p\{l, a = 2\} = \frac{u_2}{\sqrt{v_0^2 + 4v_1}} \left(t_1^{-l} - t_2^{-l} \right) \tag{57}$$

where

$$t_1 = \frac{-v_0 + \sqrt{v_0^2 + 4v_1}}{2v_1}, \quad t_2 = \frac{-v_0 - \sqrt{v_0^2 + 4v_1}}{2v_1}$$

v_0 is the that after one error symbol another error symbol follows, where the distance between the two error symbols is zero ($k = 0$). That is, the sequence described as "$\mathcal{LL}$" (for such a double error). v_1 is the probability that after an error symbol, a gap of length k $= 1$ follows, the sequence "$\mathcal{OL}$". For the A-Model, $u(0) = 1$ see (19), (25) in part I.

$$v_0 = v(0) = 1 - \alpha C,$$

$$v_1 = \alpha \left(1 + \frac{1 + \alpha}{2} C \right) C$$

Example 4

In A-Model Example 1, the probability was calculated that a burst had the length $l = 5$, where no more then one error-free symbol occurs between two symbol errors in the burst!

Solution: The burst is finished when two error-free steps occur consecutively

$$v_0 = 1 - 0.84 \cdot C = 0.16019 \quad \text{where } C = 0.99977929$$

$$v_1 = 0.84 \cdot \left(1 - \frac{1 + 0.84}{2} C\right) \; C = 0.0606736$$

$$t_{1/2} = \frac{-0.1601186 \pm 0.543215}{0.131472}; \qquad t_1 = +2.84333; \quad t2 = -5.22166;$$

$$u(2) = \frac{\alpha(1 + \alpha)}{2} C^2 = 0.77246;$$

$$p\{l, a = 2\} = 1.42201 \left[2.84333^{-l} - (-5.22153)^{-l}\right].$$

Intuitively, $p\{l = 1, a = 2\} = u(2)$. For $l = 5$, we obtain

$$p\{l = 5, a = 2\} = 1.42201 \left[2.84333^{-5} - (-5.22153)^{-5}\right],$$
$$p\{l = 5, a = 2\} = 0.0080182.$$

In a sample with 9807 errors (Example 1), there are 7577 bursts, when $a = 2$ is selected, whereby 61 bursts have the length $l = 5$. Fig. 2.4 shows the probabilities for the burst length l. It is shown that the alternating portion of t_2 in (57) disappears quickly, so that the function for longer lengths is linear when given semi-logarithmically.

However, if the distance parameter $a \gg 1$, the probability for burst lengths with large values for l can only be estimated asymptotically. Considering the general approach (46), and taking into account

$$\sum_{k=1}^{\infty} v_k = 1$$

$$t_2 = t_1 - \frac{f(t)}{f'(t)}$$

it can be seen that if an x-intercept t_0 nears 1, the greater will be the distance parameter a. With the help of the approximation method from Newton, it is possible for large values of a, that these x-intercepts can be estimated recursively.

$$t_{01} = 1; \quad t_{02} = t_{01} - \frac{1 - v_0 \, t_{01} - v_1 \, t_{01}^2 - \cdots - v_{a-1} \, t_{01}^a}{v0 + 2v_1 \, t_{01} + 3v_2 \, t_{01}^3 + \cdots + a \, v_{a-1} \, t_{01}^1}. \tag{58}$$

This results in an asymptotic estimate of the burst length distribution of (46), which can then be written as:

$$B_a(t) = \frac{u(a)}{\chi(t)}. \tag{59}$$

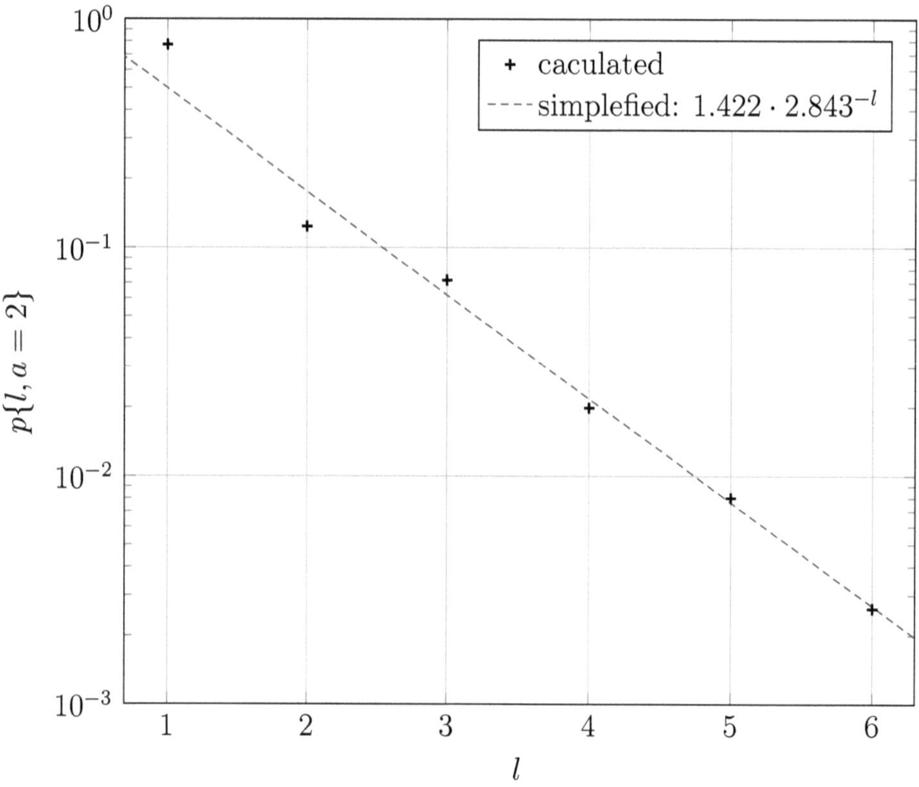

Figure 2.4.: Probability for the burst length l in bursts where $a = 2$,
A-Model: $\alpha = 0.84; p_\mathrm{e} = 8.488 \cdot 10^{-4}$

$B_a(t)$ has a simple x-intercept t_0, which is smaller than all other x-intercepts. The related residual is calculated from:

$$\operatorname*{Res}_{t=t_0} \{B_a(t)\} = \frac{u(a)}{\chi'(t_0)} \, t^{-(s+1)}. \tag{60}$$

Thereby an asymptotic estimate is obtained for the burst length distribution by ignoring the other residuals, analog (56)

$$p\{l, a\} = \frac{u(a)\, t_0^{-l}}{v_0 + 2v_1\, t_0 + 3v_2\, t_{02} + \cdots + a\, v_{a-1}\, t_0^{a-1}}$$

$$p\{l, a\} \approx \frac{u(a)\, t_0^{-l}}{\sum_{k=0}^{a-1}(k+1)\, v_k\, t_0^k}. \tag{61}$$

where

$$a = \text{burst distance parameter}$$
$$v_k = \text{probability } v(k) \text{ of the gap length } k$$
$$t_0 = \text{x-intercept from (58)}$$

2.4. Distribution of Bursts with length l and weight g

When using block-free code, it is necessary to estimate the reduction factor for symbol errors. The reduction factor is understood to be the ratio of undetected symbol errors to the total number of symbol errors, regardless of coding. From the properties of the code used, we can calculate how many bursts of length l containing g symbol errors cannot be detected.

The number of undetected symbol errors relates to the number of different kinds of bursts having the features l and g [3].

The resulting quotient is determined solely by the code properties and its value is independent of the characteristics of the disturbed channel. In the disturbed channel with bursts of features l and g there is the probability that such features occur in a finite test period. Every time such a burst is undetected by the code, then g symbol errors remain unrecognized.

Therefore, an approach for calculating the probability is derived as follows, where the bursts are of weight g and the length l. The distance parameter a, separating two bursts, is predetermined by the characteristics of the code.

$v(k)$ is the probability that after the first symbol error and k error-free steps, the next symbol error will occur. k must always be smaller than a, because otherwise the burst would be finished. This yields the truncated generating function for the next error in $s = k + 1$ steps.

$$F(t) = \sum_{s=1}^{a} p_s \, t^s = \sum_{k=0}^{a-1} v(k) \, t^{k+1}$$

If in s steps $g - 1$ errors occur, this results in the generating function

$$[F(t)]^{g-1} = \left[\sum_{k=0}^{a-1} v(k) \, t^{k+1} \right]^{g-1}.$$

We can describe this generating function by multiplying a power series:

$$\begin{aligned}
[F(t)]^{g-1} = \quad & q(l = g/g; a) \, t^{g-1} \\
& + q(l = g + 1/g; a) \, t^g \\
& + \dots \\
& + q(l = a(g-1) + 1/g; a) \, t^{a(g-1)}.
\end{aligned} \tag{62}$$

The coefficient q of t^x is then the probability of length $l = x - 1$, whereby $g \le l \le a(g-1) + 1$.

Then q should be multiplied with the probability $u(a)$, since after the last symbol error a gap of $k \ge a$ must occur. Then dividing q by the probability $p\{g\}$ within the set of all bursts of weight g and distance parameter a, according to (39), we find based on the conditional probability, that the bursts have length l of all bursts with the two parameters weight g and distance a.

$$p\{l/g; a\} = \frac{q(l, g) \, u(a)}{u(a) \, [1 - u(a)]^{g-1}} = \frac{q(l, g)}{[1 - u(a)]^{g-1}} \tag{63}$$

The generating function for this required probability is

$$[F^*(t)]^{g-1} = \frac{1}{[1-u(a)]^{g-1}} \left[\sum_{k=0}^{a-1} v(k)\, t^{k+1}\right]^{g-1}.\qquad(64)$$

So we can represent the required probabilities $p\{l/g; a\}$ as coefficients of t^{l-1} in a power series in closed analytical form but only for a given weight g and potentiating formula (64).

Example 5

Using the Burst Analysis Matrix in Tab. 2.1 an artificially generated error sequence $(a = 5)$

a)   Calculate the number for the occurrence of all burstweights.

The number of bursts with single errors $(g = 1)$ in this sample is 3446. Then follows $u(5) = 3446/5727 = 0.5843$, and we get Num $\{g\} = u(5)\,[1 - u(5)]^{g-1} \cdot 5727$

Num $\{g\}$					g				
	1	2	3	4	5	6	7	8	> 8
calculated	3446	1372	546	217	87	35	14	5	5
randomly chosen[†]	3446	1375	528	219	108	34	12	2	3

[†] Tab. 2.1

b)   Calculate the distribution of length l in all bursts with weight $g = 2$.

It is clear that the longest burst arises, if between the three symbol errors follow each other. We then obtain $l_{\max} = 7$. By arranging (64) in powers of t, we obtain, using values $v(x)$ and α as calculated in (20),(25) and (29) with `Excel`

$$[F^*]^1 = \frac{1}{[1-u(5)]^1} \left[v(0)\,t + v(1)\,t^2 + v(2)\,t^3 + v(3)\,t^4 + v(4)\,t^5\right]^1$$

$$\text{Num}\,\{l = 2/g = 2; a = 5\} = 1372 \cdot \frac{v(0)}{[1-u(5)]^1} = 712$$

$l =$	2	3	4	5	6	
	712	282	196	118	90	calculated
	796	238	156	96	89	randomly chosen, Tab. 2.1

c)   Calculate the distribution of length l in all bursts with weight $g = 3$.

$$[F^*]^2 = \frac{1}{[1-u(5)]^2} \left[v(0)\,t + v(1)\,t^2 + v(2)\,t^3 + v(3)\,t^4 + v(4)\,t^5\right]^2$$

$$\mathrm{Num}\,\{l = 5/g = 3; a = 5\} = 528 \cdot \frac{2v(0)\,v(2) + v(1)^2}{[1 - u(5)]^2}$$

$$= 528 \cdot \frac{0.02935}{0.1729} = 89.6; \quad \text{randomly: } 85$$

The desired result $p(l/g = 3; a = 5)$ is the coefficient of t^{l-1} from the power series. It is evident that the formal solution is easily programmable for any values of g and a.

2.5. Burst Analysis Matrix

The previously calculated probabilities can be represented clearly in a two-dimensional probability table for bursts, with features length l and weight g and a given distance parameter a. This table also contains the number of measured and sorted events and is called the **Burst Analysis Matrix**.

The channelmatrix shown in Tab. 2.1 was created by sorting an artificially generated error sequence. The column totals are the number of bursts with weight g, and the line totals are the number of bursts of length l. For example, 40 bursts were registered with errors of weight $g = 4$ and length $l = 5$. As the burst-weight distribution and burst-length distribution may be calculated with (40), (46) respectively, the approach (63) is used to split the column sum of bursts where the weight g is constant, according to burst length l. When all elements in Tab. 2.1 have been calculated in this way, the results can be double-checked using the sums of the rows. The given dividing line indicates the maximum possible burst lengths.

Table 2.1.: Burst Analysis Matrix of an artificially generated error sequence ($a = 5$)

length l	weight g									$\sum l$
	1	2	3	4	5	6	7	8	9 − 12	
1	3446									3446
2		796								796
3		238	146							384
4		156	132	38						326
5		96	85	40	7					228
6		89	69	41	11	0				210
7		0	56	30	14	3	0			103
8		0	24	34	19	1	2	0		80
9 − 12		0	16	34	47	25	5	1	0	128
13 − 18		0	0	2	10	4	5	1	3	25
19 − 25		0	0	0	0	1	0	0	0	1
$\sum g$	3446	1375	528	219	108	34	12	2	3	5727

Part III.

Probability of error structures in blocks

3. Calculation of error structures in blocks

In many error control methods, the data are encoded and transmitted in blocks. A block is a group of symbols which is encoded and transmitted as a unified whole, in order to determine transmission errors. Unrecognisable error structures should occur sufficiently rarely. Therefore, it is of interest to calculate with certainty the probabilities of non-detectable error structures regardless the relevant coding.

Some introductory remarks illustrate the need for, and the potential applications of, such calculations.

In the development of various codes, the binary channel has been assumed to be "memory-less".

The resultant classical works of coding theory, for example Peterson [10], have gained wide acceptance.

However, in consideration of Tab. 3.1, it can be seen that in "memory-less" channels within disturbed blocks, almost always single errors occur, provided that the symbol error probability is $\ll 0.5$.

Therefore, codes have been developed to correct single errors. Later, however, actual measurements revealed that the proportion of multiple errors is in fact significantly higher in disturbed blocks, as suspected. We define *bundle errors* between the first and last error with length l and g errors in blocks of length n.

Cyclical codes detect bundle errors in blocks up to a certain length l. For longer bundle errors, the probability of undetected error blocks can be estimated. These codes identify further error blocks up to a weight g. Channel measurements described in the literature therefore focus on specifying the length distribution and the weight distribution of bundle errors within disturbed blocks.

Table 3.1.: Measured and calculated weight distributions in disturbed blocks

weight g	measured[†]	calculated[‡]
1	2373	3191
2	294	1.649
3	54	$4.26 \cdot 10^{-4}$
4	11	$5.5 \cdot 10^{-8}$
5	4	$2.8 \cdot 10^{-12}$

Example with $n = 5$ bits and $p_{\mathrm{e}} = 2.58430 \cdot 10^{-4}$

[†] UHF 600 bit/s; [‡] based on Bernoulli

Table 3.2.: Channel matrix of an artificially generated error sequence for bundle errors of length l and weight g in blocks of length $n = 15$

length l	weight g										$\sum l$
	1	2	3	4	5	6	7	8	$9-12$	$13-15$	
1	3469										3469
2		750									750
3		214	141								355
4		133	128	32							293
5		77	76	36	6						195
6		67	59	30	11	1					168
7		55	42	21	8	2	0				128
8		37	33	25	12	3	1	0			111
$9-12$		74	88	54	33	14	2	2	0		267
$13-15$		19	19	17	8	5	2	0	0	0	70
$\sum g$	3469	1426	586	215	78	25	5	2	0	0	5806

$$\overset{\longleftarrow a \longrightarrow}{} \quad \boxed{\mathcal{L}} \quad \overset{\longleftarrow b \longrightarrow}{}$$
$$\longleftarrow n = a + b + 1 \longrightarrow$$

Figure 3.1.: Calculation of the probability of single errors in a block of length n

It has been noted that there are particular short bundle errors with certain weights which remain particularly frequently undetected by the code. Therefore, it is proposed as in [9], to sort the bundle error occurring in blocks according to the two features length l and weight g, and to tabulate them by way of a so-called channel matrix (Tab. 3.2), as was done in Tab. 2.1.

Not all coding theorists or developers of transmission equipment have had access to results of error structure measurements. Therefore in the following section probability formulas are derived for the occurrence of elements in this channel matrix and in several illustrative examples, based on the A-Model.

3.1. Single errors in blocks of length n

A single symbol error can be located in each of the positions of the block (Fig. 3.1). The required error probability is obtained as followed:

$$p\{n, g = 1\} = p_e \begin{bmatrix} u(0)\, u(n-1) \\ +u(1)\, u(n-2) \\ \cdots \\ +u(n-1)\, u(0) \end{bmatrix}. \tag{65}$$

The convolution of the probability distribution $u(k)$, as per (65), has the generating function $[U(t)]$

$$\text{Given } U(t) = (1 - C\,t)^{-\alpha}, \tag{66}$$

$$\text{then } [U(t)]^2 = (1 - C\,t)^{-2\alpha}. \tag{67}$$

Taking into account (19) from Part I, for n=2,3,4,...

$$p\{n, g = 1\} = p_{\mathrm{e}}\,\frac{\Gamma(2\alpha + n - 1)}{\Gamma(n)\,\Gamma(2\alpha)}\,C^{n-1}$$

$$p\{n, g = 1\} = p_{\mathrm{e}}\,\frac{2\alpha(2\alpha + 1)\cdots(2\alpha + n - 2)}{(n - 1)!}\,C^{n-1} \quad \text{where } p\{n = 1, g = 1\} = p_{\mathrm{e}}. \tag{68}$$

$\Gamma(\alpha)$ is calculated with the help of `Excel` as *"=Exp(GammaLn(α))"*.

An extreme case for strongly bundled channels exists: If the bundling factor $1 - \alpha$ reaches the value 0.5, then $2\alpha = 1$ and $\Gamma(1) = 1$ and we have

$$p\{n, g = 1, \alpha = 0.5\} = p_{\mathrm{e}}\,C^{n-1} = p_{\mathrm{e}}\,(1 - p_{\mathrm{e}})^{n-1}$$

Another extreme case scenario arises when $\alpha = 1$, i.e. for a channel without memory. By using (68), whereby $\Gamma(2) = 1$ we obtain

$$p\{n, g = 1\} = n\,p_{\mathrm{e}}\,C^{(n-1)} \approx n\,p_{\mathrm{e}}\,(1 - p_{\mathrm{e}})^{n-1}$$

for the probability of individual errors in blocks of length n for channels without memory. For large values of n according to (22), Part I, we obtain

$$p\{n, g = 1\} = \frac{p_{\mathrm{e}}}{\Gamma(2\alpha)}\,(n - 1)^{2\alpha - 1}\,C^{n-1}. \tag{69}$$

Example 6

What is the probability that a single error occurs when we have a block length $n = 1023$ or the block length $n = 5$? Use the A-Model again as per the measured sample in Example 4 whereby $p_{\mathrm{e}} = 0.0008488$ and $\alpha = 0.84$, so is $C = 0.999779$, the number of transmitted symbols is 11,554,541 and the number of symbol errors $z_{\mathrm{f}} = 9807$.

Solution $(n = 1023)$: From (69)

$$p\{n, g = 1\} = \frac{p_{\mathrm{e}}}{\Gamma(2\alpha)}\,(n - 1)^{2\alpha - 1}\,C^{n-1}$$

we get with `Excel`:$\Gamma(x) = Exp(GammaLn(1.68))$

$$p\{n, g = 1\} = \frac{p_{\mathrm{e}}}{\Gamma(1.68)}\,(1023 - 1)^{0.68}\,C^{1022} = \frac{0.0008488}{0.905}\cdot 111.2824\cdot 0.79781 = 0.083327$$

From $11.554.541/1023 = 11.295$ transferred blocks, each of 941 of these blocks should contain a single error. However, in all, 1096 blocks containing a single error were actually detected in our sample.

Solution $(n = 5)$:   We obtain from (68)

$$p\{n, g = 1\} = p_{\mathrm{e}} \frac{2\alpha\,(2\alpha + 1)\dots(2\alpha + n - 2)}{(n - 1)!}\,C^{n-1}$$

$$p\{5, g = 1\} = p_{\mathrm{e}} \frac{1.68 \cdot 2.68 \cdot 3.68 \cdot 4.682}{4!}\,0.99912 = 0.00274$$

From $11,554,541/5 = 2,310,908$ transferred blocks, $6,332$ blocks should therefore theoretically each contain a single error. In practice, $6,271$ blocks were found to each contain a single error in our sample.

This demonstrates how the A-Model provides good estimates.

Using a single error correction code, 10% of symbol errors will be sure to be corrected for a block length of $n = 1023$ bits. However, 64% of symbol errors will be corrected for a block length of $n = 5$ bits.

3.2. Distribution of error length l in errorneus blocks

Different codes have the capacity to detect errors in disturbed blocks provided the length l of the error bundle in the block does not exceed a certain value. Such a bundle error is shown in Fig. 3.2. The burst begins and ends with a symbol error. The intervening steps are either erroneous or errorfree. If the error length is the same as the block length, i.e. $l = n$ and $s = n - 1$, then the error probability is $p\{l, n\}$ for an error bundle of length n:

$$p\{l, n\} = p\{l = n, n\} = u(0)\,r(n - 1)\,u(0),$$
$$p\{l = n, n\} = r(n - 1).$$

$r(n - 1)$ is here the symbol correlation probability (27), as in Part I.

The probability that in a block of length n an error bundle of length $l = n$ is included:

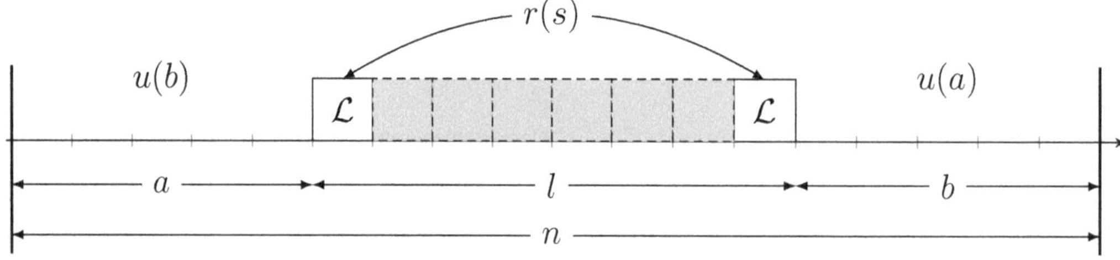

Figure 3.2.: Error bundle of length l in a block of length n

$$p\{l = n, n\} = p_{\mathrm{e}}\left(1 - \frac{\alpha}{1!}C^1 - \frac{\alpha\,(1-\alpha)}{2!}C^2 - \cdots - \frac{\alpha\,(1-\alpha)\ldots(n-2-\alpha)}{(n-1)!}C^{n-1}\right)$$

$$(70)$$

It then follows

$$p\{l = 1, n = 1\} = p_{\mathrm{e}},$$
$$p\{l = 2, n = 2\} = p_{\mathrm{e}}\,(1 - \alpha\,C).$$

For the general case $l \leq n$, the required probability $p\{l, n\}$ comprises two factors. The first factor p_1 is the probability that the number of error-free symbols occurring before and after the error bundle of length l in the block is the sum $a + b$ (see Fig. 3.1). The corresponding probability is given by an illustrative comparison of Fig. 3.1 with Fig. 2.4 and (68), as a single error probability in a block of length n.

$$n^* = a + b + 1 = n - s = n - l + 1$$
$$p_1 = p\{n^*, g = 1\}$$

$$p\{n, g = 1\} = p_{\mathrm{e}}\frac{2\alpha\,(2\alpha + 1)\cdots(2\alpha + n - 2)}{(n - 1!)}C^{n-1} \qquad (71)$$

The second factor p_2 is the probability that when one error occurs, another error will occur after exactly $s = l - 1$ bits (27), as in Part I.

$$p_2 = \frac{r(2)}{p_{\mathrm{e}}}.$$

From this the probability follows, that an error bundle of length l occurs in a block of length n:

$$p\{l, n\} = p_{\mathrm{e}}\frac{2\alpha(\alpha + 1)\cdots(2\alpha + n - l - 1)}{(n - 1)!}C^{n-l}\frac{r(l - 1)}{p_{\mathrm{e}}} \qquad (72)$$

where

$$\frac{r(l - 1)}{p_{\mathrm{e}}} = 1 - \frac{\alpha}{1!}C - \frac{\alpha\,(1-\alpha)}{2!}C^2 - \cdots - \frac{\alpha\,(1-\alpha)\cdots(l-2-\alpha)}{(l-1)!}C^{l-1}$$

for $l = 3, 4, \ldots$ and

$$\frac{r(l - 1)}{p_{\mathrm{e}}} = 1 - \frac{\alpha}{1!}C \quad \text{for } l = 2$$

or written differently for $l = 3, 4, \ldots$:

$$p\{l, n\} = p_{\mathrm{e}}\,A\,B \text{ with} \qquad (73)$$
$$A = \frac{2\alpha(\alpha + 1)\cdots(2\alpha + n - l - 1)}{(n - l)!}C^{n-1},$$
$$B = 1 - \frac{\alpha}{1!}e^{-\beta} - \frac{\alpha(1-\alpha)}{2!}C^2 - \cdots - \frac{\alpha\,(1-\alpha)\ldots(l-2-\alpha)}{(l-1)!}C^{l-1}.$$

Example 7

Calculate the probability for the proportion of error bundles with a length $l > 3$ in disturbed blocks of length $n = 5$ for the A-Model according to Example 1.

Solution: Where $\alpha = 0.84$, $p_\mathrm{e} = 0.0008488$ and $\beta = \frac{1}{4486}$ is $C = \left(1 - p_\mathrm{e}^{\frac{1}{\alpha}}\right) = 0.99978$

$$p_\mathrm{b}(5) \approx p_\mathrm{e}\, n^\alpha = 0.00328 \quad \text{or more precisely:}$$

$$p_\mathrm{b}(5) = p_\mathrm{e} \sum_{k=0}^{4} u(k).$$

Recursively it follows from (20), in Part I

$$u(0) = 1; \quad u(1) = \alpha\, C; \quad u(2) = u(1)\,\frac{1+\alpha}{2}\, C$$

$$u(k) = u(k-1)\frac{k-1+\alpha}{k}\, C, \quad p_\mathrm{b}(5) = 0.003435$$

However a block error frequency of 0.0033926 was measured in a finite sample.

According to Example 6, the probability of a single error is $p\,\{5, g = 1\} = 0.00274485$. From (73), it follows:

$$p\,\{l = 2, n = 5\} = pe\, \frac{1.68 \cdot 2.68 \cdot 3.68}{3!}\, C^3\,(1 - 0.84\,C),$$

$$= 0.000375.$$

$$p\,\{l = 3, n = 5\} = pe\, \frac{1.68 \cdot 2.68}{2!}\, C^2\left(1 - 0.84\,C - \frac{0.84 \cdot 0.16}{2}\, C^2\right)$$

$$= 0.0001773.$$

Thus, the proportion of blocks with error bursts $l > 3$ in disturbed blocks of length $n = 5$ is

$$\frac{p\,\{l > 3, n = 5\}}{p_\mathrm{b}(5)} = 1 - \frac{0.00274485 + 0.000375 + 0.0001773}{0.003435} = 0.040131.$$

Only 4% of erroneous blocks with a block length $n = 5$ have error bundles of length $l > 3$.

This corresponds to just 319 of all $2{,}310{,}908 \cdot 0.003435 = 7938$ erroneous blocks. In the measured sample of 7840 blocks, 426 blocks were found to have error bursts have error bundles of length $l > 3$.

It is seen that in the A-Model where there are independent gaps between symbol errors, longer error bundles occur less frequently than in the measured sample. Real

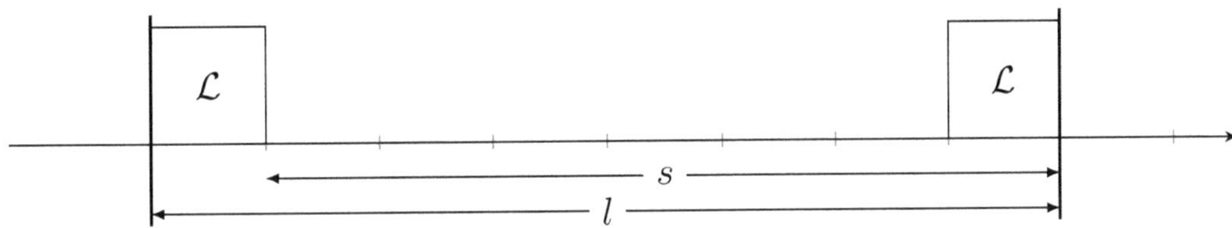

Figure 3.3.: Calculating the distribution of weight in error bundles of length l

samples reveal that after one short gap another short gap is more likely to follow than a large gap. This memory feature need only be considered in more complicated models [3]. Tab. 3.3 shows the comparison of measured and calculated values.

3.3. Distribution of weight g in error bundles of length l

An error bundle within a block differs from a burst therein, when the number of consecutive error-free bits is not limited within the error bundle. However, in contrast no more than $a-1$ error-free bits are allowed to follow each other in the burst, otherwise it will be terminated. If in a disturbed block an error bundle of length l occurs, then for the purpose of calculating the immunity of block codes the number g (weight) of symbol errors in the error bundle [3], [9] must be known. To solve this problem using the A-Model for disturbed digital channels, the probability should be calculated so that after an erroneous bit in s steps exactly $g-1$ errors occur, Fig. 3.3. The error structure to be considered contains exactly g errors within the interval of length l, whereby the first error has already occurred. The last error occurs in exactly the last position.

$[P(t)]^{g-1}$ is the generating function for the probability of $g-1$ erroneous symbols occurring in exactly s steps. The desired probabilities arise as the coefficients of the

Table 3.3.: Length l of error bundles in block of length $g = 5$

Length l of error bundles in block of length 5	Number of measured blocks	Calculated with the A-Model
1	6271	6345
2	680	861
3	463	408
4	266	220
5	160	104
Summe	7840	7938

power series expansion of

$$[P(t)] \quad \text{at the point } t = 0$$

$$[P(t)] = \sum_{s=0}^{\infty} p(s, g)\, t^s. \tag{74}$$

Using (25) from Part I and $P(t) = t\,V(t)$, we have

$$[P(t)]^{g-1} = \left[1 + \frac{t-1}{(1 - C\,t)^{\alpha}} \right]^{g-1} = (1 + y)^{g+1}. \tag{75}$$

According to the binomial theorem, we can write:

$$(1 + y)^{g-1} = 1 + (g - 1)\,y + \binom{g-1}{2}\,y^2 + \binom{g-1}{3}\,y^3 + \cdots + \binom{g-1}{g-1}\,y^{g-1} \tag{76}$$

$$\text{with } y^r = (t - 1)^r\,(1 - C\,t)^{\alpha r}.$$

Then we substitute in $u = (t - 1)^r$ and $v = (1 - C\,t)^{-\alpha r}$ to give $y^r = u\,v$. For the i-th derivative of a product, we can write:

$$y^r = \binom{i}{0}\overset{(i)}{u}\,v + \binom{i}{1}\overset{(i-1)}{u}\overset{(1)}{v} + \binom{i}{2}\overset{(i-2)}{u}\overset{(2)}{v} + \cdots + \binom{i}{i-1}\overset{(1)}{u}\overset{(i-1)}{v} + \binom{i}{i}\,u\,\overset{(i)}{v}. \tag{77}$$

For the i-th derivative of u at the point $t = 0$, it then follows

$$\overset{(i)}{u}(0) = (-1)^{r-i}\,r(r - 1)(r - 2)\ldots(r - i + 1) = (-1)^{r-i}\binom{r}{i}\,i! \tag{78}$$

with $\overset{(i)}{u}(0) = 0$ for $i > r$, i.e. there are only the first r derivations. For the i-th derivative of v at the point $t = 0$, it follows:

$$\overset{(i)}{v}(0) = \alpha r\,(\alpha r + 1)(\alpha r + 2)\ldots(\alpha r + i - 1)\,C^i, \tag{79}$$

$$\overset{(i)}{v}(0) = C^i \prod_{k=1}^{i}(\alpha r + k - 1) \tag{80}$$

where

$$\prod_{k=1}^{j} a_k = a_1 a_2 a_3 a_j. \tag{81}$$

From (74), the desired probabilities $p(s, g)$ are calculated by adding all coefficients t^s from the power series expansions of the summands from the binomial (76) at the point $t = 0$:

$$p(s, g) = \binom{g-1}{1}\frac{[y(0)]^{(s)}}{s!} + \binom{g-1}{2}\frac{[y^2(0)]^{(s)}}{s!} + \cdots + \binom{g-1}{g-1}\frac{[y^{g-1}(0)]^{(s)}}{s!}, \tag{82}$$

$$p(s, g) = \frac{1}{s!}\sum_{r=1}^{g-1}\binom{g-1}{r}[y^r(0)]^{(s)}. \tag{83}$$

By substituting (78) and (80) into (77), we have

$$[y^r(0)]^{(s)} = \sum_{j=0}^{s} \binom{s}{j} [u(0)]^{(s-j)} [v(0)]^{(j)}$$

$$= \sum_{j=s-r}^{s} \binom{s}{j} (-1)^{r-s+j} \binom{r}{s-j} (s-j)! \, C^j \prod_{v=1}^{j} (\alpha r + v - 1). \tag{84}$$

And finally, by inserting the formula (84) into (83), we have

$$p(s,g) = \sum_{r=1}^{g-1} \frac{1}{s!} \binom{g-1}{r} \left[\sum_{j=s-r}^{s} (-1)^{j+r-s} \binom{s}{j} \binom{r}{s-j} (s-j)! \, C^j \prod_{v=1}^{j} (\alpha r + v - 1) \right]. \tag{85}$$

Taking into account

$$\binom{s}{j} = \frac{s!}{j!\,(s-j)!}$$

we can summarize as follows

$$\frac{1}{s!}(s-j)! \binom{s}{j} = \frac{1}{j!}.$$

For simplification, the functions of type

$$u(r,j) = \frac{C^j}{j!} \prod_{v=1}^{j} (\alpha\, r + v - 1) \tag{86}$$

where $u(1,k) = u(k)$, $u(r,0) = u(0) = 1$ are introduced, since they can be treated mathematically as the function $u(k)$ with all approximations, see (69).

These functions are referred to as **distance functions**. Given $l = s + 1$, the conditional probability $p(l,g)$, that the error structure of length l is started with a symbol error, contains exactly g erroneous symbols, i.e. that the weight is g, is then:

$$p(l,g) = \sum_{r=1}^{g-1} \sum_{j=l-1-r}^{l-1} (-1)^{j+r+1-l} \binom{g-1}{r} \binom{r}{l-1-j} u(r,j), \tag{87}$$

where $l \geq g$ and $g = 2, 3, \ldots$,

$$u(r,j) = \frac{\alpha r \,(\alpha r + 1) \ldots (\alpha r + j - 1)}{j!} \, C^j,$$

$$u(1,k) = u(k),$$

$$u(r,0) = u(0) = 1.$$

Here, the error patterns occur before and after the interval of length l regardless of the result obtained from (87).

The number of summands in this equation is

$$S_g = \frac{g-1}{2}(2+g).$$

This number is independent of the length of the error pattern. There are $\binom{l-2}{g-2}$ error patterns of length l with weight g where each pattern is a unique structure. For instance, we want to find the probability $p(l=18, g=10)$. Then we must calculate $\binom{16}{8} = 12870$ individual single error pattern probabilities. However, using formula (87), we simply need to calculate $\frac{1}{2}(g-1)(2+g) = 54\,$summands. Thus using formula (87), considerable calculation time is saved.

Example 8

Using the A-Model with the sample in Example 1, calculate the probability that in a block of length $n = 5$, the first and the last symbols are incorrect and a total of 3 errors occurs!

Solution: The first error occurs in position 1 with the probability p_e (”history” unknown). The second error can be located at any of positions 2, 3 or 4, whereas the last error occurs in the 5th position. Therefore, three different single patterns must be calculated. Thus follows:

$$p\{M\} = p_e(v_0v_2 + v_1v_1 + v_2v_0) = p_e\left(v_1^2 + 2v_0v_2\right) = p_e \cdot 0.01778707 \qquad (88)$$

For larger error bundles, using this formula would require us to calculate far too many summands. Therefore, the solution is best given by using formula (87): $p\{M\} = p_e\, p(l=5, g=3)$.

Calculation scheme for (87):

r	j	$\hat{a}_{r,j}\, u(r,j)$	r	j	$\hat{a}_{r,j}\, u(r,j)$
1	3:	$+\binom{2}{1}\binom{1}{1} u(1,3)$	2	2:	$+\binom{2}{2}\binom{2}{2} u(2,2)$
1	4:	$-\binom{2}{1}\binom{1}{0} u(1,4)$	2	3:	$-\binom{2}{2}\binom{2}{1} u(2,3)$
			2	4:	$+\binom{2}{2}\binom{2}{0} u(2,4)$

$$p\{M\} = p_{\mathrm{e}}\left[2u(1,3) - 2u(1,4) + u(2,2) - 2u(2,3) + u(2,4)\right]. \qquad (89)$$

In contrast to the elementary calculation from (88), only 5 distribution functions $u(r,j)$ need to be determined. However, the approach (87) for long error bundles proves indispensable.

$$+2u(1,3) = \frac{2}{3!}\,\alpha(\alpha+1)(\alpha+2)\,C^3 = 2u(3) \qquad = +1.46219942$$

$$-2u(1,4) = -\frac{2}{4!}\,\alpha(\alpha+1)(\alpha+2)(\alpha+3)\,C^4 = -2u(4) = -1.403401634$$

$$+u(2,2) = \frac{2}{2!}\,\alpha(2\alpha+1)\,C^2 \qquad = +2.250206399$$

$$-2u(2,3) = -\frac{2}{3!}\,2\alpha(2\alpha+1)(2\alpha+2)\,C^3 \qquad = -5.519287951$$

$$+u(2,4) = \frac{2}{4!}\,\alpha(2\alpha+1)(2\alpha+2)(2\alpha+3)\,C^4 \qquad = +3.228070837$$

$$= \underline{0.01778707}$$

where $\alpha = 0.84$, $p_{\mathrm{e}} = 0.0008488$ and $C = \left(1 - p_{\mathrm{e}}^{\frac{1}{\alpha}}\right) = 0.999779239$ according to Example 1.

When an error occurs in position 1 as in well as position 5, the probability that two further errors occur within the three interim positions is 1.78%. Considering that the likelihood that the first symbol error occurs at position 1 is only $p_{\mathrm{e}} = 0.0008488$, we can see the "memory" of the channel as reflected in the A-Model. In the sample from Example 1 with 11,554,541 transmitted bits,

$$z_{\mathrm{M}} = \frac{11554541}{5}\, p_{\mathrm{e}}\, p(l = 5, g = 3) = 35$$

error patterns of this kind must occur. 45 such patterns were measured. If, however, the normal distribution for the measured events is to be believed (based on large trial numbers), then event numbers will only occur outside the 3σ boundaries with a so-called error probability of $\varepsilon = 0.0027$. See part I, Koller [11]. The normal distribution assumed; a deviation of

$$\Delta z \approx \pm 3\sqrt{p\,(1-p)\,n}$$

in the measurement is allowed. With $1 - p \approx 1$ and $n\,p = 35$, we get $\Delta z \approx \pm 17$. The resulting calculation will not be rejected with an error probability of 0.27% for the real channel, as long as 35 ± 17 events have been measured.

3.4. Distribution of weight g of error bundles of length l into erroneous blocks

If we divide the error bundles occurring in erroneous blocks into classes, we can determine the proportion of error bundles not recognizable by an error-protection code within each class. We can assign the appropriate weight to each probability of occurrence per class, and determine the average over all classes. The result is the probability of the occurrence of blocks that contain undetected error bundles despite having an error-protection code. This important result can be divided between calculations for the channel side and calculations for the code side. The probability of an error bundle class occurring depends on the channel characteristics and is calculated as follows. The calculation depends on the proportion of undetectable error bundles of the code used. See [12], [3], [9].

Fig. 3.4 represents how we calculate the probability bundle occurs of length $p\,\{l, g, n\}$, that in a block of length n an error l with g symbol errors. In our calculation we must note that there are $n - l + 1$ different distances of the first error symbol from the beginning of the block. In addition, $\begin{pmatrix} l - 2 \\ g - 2 \end{pmatrix}$ combinations are possible for ordering $g - 2$ internal symbol errors within the error bundle. We need therefore to find the class probability for z_M single patterns:

$$z_\mathrm{M} = (n - l + 1) \begin{pmatrix} l - 2 \\ g - 2 \end{pmatrix}. \tag{90}$$

There are for example, for the values $n = 63, l = 5, g = 3$, exactly 177 different patterns.

$$z_\mathrm{M} = 59 \begin{pmatrix} 3 \\ 1 \end{pmatrix} = 177.$$

For the class probability, using the same approach as in (71) and (87) for the A-

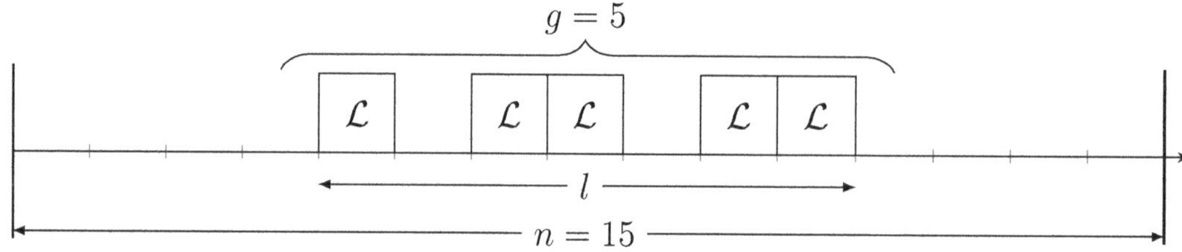

Figure 3.4.: Calculating the probability of the occurrence of error bundles of the length l with g symbol errors in block of length n

Model, we obtain:

$$p\{l, g, n\} = p_{\mathrm{e}} \frac{\Gamma(2\alpha + n - \alpha)}{\Gamma(1 + n - l)\Gamma(2\alpha)} C^{n-l} p(l, g). \tag{91}$$

The first factor is rewritten as the analog formula (68)

$$p_{\mathrm{e}} \frac{\Gamma(2\alpha + n - \alpha)}{\Gamma(1 + n - l)\Gamma(2\alpha)} C^{n-l} == p_{\mathrm{e}} \frac{2\alpha (2\alpha + 1) \dots (2\alpha + n - l - 1)}{(n - l)!} C^{n-l}$$

With the distance function $u(r, j)$ as defined in (86), it follows from (87) and (91), for $g = 2, 3, \dots$ and $2 \le g \le l \le n$

$$p\{l, g, n\} = p_{\mathrm{e}}\, u(2, n-l) \sum_{r=1}^{g-1} \sum_{j=l-1-r}^{l-1} (-1)^{j+r+1-l} \binom{g-1}{r} \binom{r}{l-1-j} u(r, j) \tag{92}$$

This gives us the probability that in a block of length n error bundles of length l with g errors occur.

This expression is simply the product of the probabilities that in a block of length $n - l + 1$ a single error occurs, and that following this single error, an error bundle of length l with a total of g symbol errors occurs. For example, for the values $n = 63, l = 5, g = 3$, all 177 possible probabilities using the single pattern (90) are replaced by 5 summands. This equation is a principal **Channel Equation** based on a perspective of coding theory, for the probability of an error bundle class of length l and g symbol errors occurring in blocks of length n.

Example 9

Calculate the probability $p\{l, g, n\}$, for error bundles of length $l = 5$ with $g = 3$ symbol errors occurring in blocks of length $n = 63$. Here, the A-Model is used according to Example 1.

Solution: For the factor $u(2, n - l)$, the approximation (69) applies for large values of $n - l$.

$$u(2, n - l) \approx \frac{1}{\Gamma(2\alpha)}(n - l)^{2\alpha - 1} C^{n-l}, \quad \text{whereby } C = (1 - p_{\mathrm{e}}^{\frac{1}{\alpha}}) = 0.999779 \tag{93}$$

$$u(2, 58) \approx \frac{1}{\Gamma(1.68)} \cdot 58^{0.68}\, C^{58},$$

$$u(2, 58) \approx 1.11279015 \cdot 15.8172521 \frac{1}{1.01290} = 17.3770$$

With the help of **Excel**: $\Gamma(\alpha) = \mathrm{Exp}(\mathrm{GammaLn}(\alpha))$

The second factor was calculated in Example 8 as $p(l = 5, g = 3) = 0.017788$. With $p_{\mathrm{e}} = 0.0008488$ (from Example 8), we obtain the desired probability:

$$p\{l = 5, g = 3, n = 63\} = p_{\mathrm{e}} \cdot 17.2535 \cdot 0.017788 = p_{\mathrm{e}} \cdot 0.30690.$$

Thus, $z_\mathrm{M} = 48$ such patterns.

$$z_\mathrm{M} = \frac{11,554,541}{63} \cdot p_\mathrm{e} \cdot 0.3069 = 48;\ z_\mathrm{M} \approx \pm 3\sqrt{48} = \pm 21$$

must be included in the sample with 11,554,541 transmitted symbols. However, only 18 patterns of this type were registered. This deviation is not random, but rather it refers to the tendency, that in the A-Model, assuming independent gaps, short errors bundles occur more frequently while long error bundles are rarer than is the case with real channels.

Burst process models (HMM) are by definition more accurate. The construction of such burst process models (HMM) is described by Proakis [13], Turin [14] and Wilhelm [12], [15], [16], [3].

3.5. Distribution of weight g of error bundles in erroneous blocks (weight spectrum)

In coding theory, the channel was described initially by specifying the distribution of the number of symbol errors in erroneous blocks.

For n different classes, the probabilities $p\{g, n\}$ can be calculated using formula (92).

$$p\{g, n\} = \sum_{l=g}^{n} p\{l, g, n\}, \tag{94}$$

$$p\{g, n\} = p_\mathrm{e} \sum_{l=g}^{n} \left[u(2, n-l) \sum_{r=1}^{g-1} \sum_{j=l-1-r}^{l-1} (-1)^{j+r+1-l} \binom{g-1}{r} \binom{r}{l-1-j} u(r, j) \right] \tag{95}$$

$$u(r, j) = \frac{C^j}{j!} \prod_{m=1}^{j} (\alpha r + m - 1);\ u(1, k) = u(k);\ u(r, 0) = u(0) = 1 \quad \text{for } g = 2, 3, \ldots$$

This probability for the weight g in disturbed blocks of length n is the sum of all the probabilities with g errors over all lengths l. This formula is particularly suitable for programmed calculations. This probability contains the value of summand as follows:

$$\text{Value} = \frac{g-1}{2}(2+g)(n-g+1)$$

Thereby $\binom{n}{g}$ different error structures can be calculated. For $n = 63, g = 10$, we obtain $\binom{n}{g} \approx 1.278 \cdot 10^{11}$. Alternatively, using the above formula simply:

$$\frac{g-1}{2}(2+g)(n-g+1) = 2916.$$

Using formula (95) with fast computers yields calculations in a matter of seconds.

Example 10

Calculate the probability $p\{g, n\}$ for all error bundles with $g = 2$ symbol errors occurring in blocks of length $n = 3$ using formula (95).

Solution:

$$p\{g = 2, n = 3\}\,/p_{\mathrm{e}} = u(2,1)\,[u(1,0) - u(1,1)] + u(2,0)\,[u(1,1) - u(1,2)]$$
$$= 2\alpha\,C\,v(0) + v(1) = 2\,u(1)\,v(0) + v(1)$$

This is plausible because there are three different pattern exists:

$$\mathcal{LLO} \mathrel{\hat=} v(0)\,u(1); \quad \mathcal{LOL} \mathrel{\hat=} v(1); \quad \mathcal{OLL} \mathrel{\hat=} v(0)\,u(1)$$

4. Appendix – Basic formulars for the A-Model

4.1. Definitions

Limiting factor for convergence to the BSC-Channel without memory

$$C = \mathrm{e}^{-\beta} \equiv \left(1 - p_{\mathrm{e}}^{\frac{1}{\alpha}}\right) \tag{24}$$

Error distance

$$s = k + 1$$

Number of bursts in a sample

$$z_{\mathrm{b}} = z_{\mathrm{f}}\, u(a) \tag{37}$$

4.2. General relationsships for the A-Model

Symbol error probabilit y as a function of the error distance, (3.107) in [3]

$$p_{\mathrm{e}}(k) = 1 - \frac{k + \alpha}{k + 1}\, C$$

Block error probability

$$p_{\mathrm{b}}^{(n)} = p_{\mathrm{e}}\, n^{\alpha} \text{ for } n < n^{*} \tag{2}$$

Probability of $\geq k$ error free symbols after symbol error, for $k = 1, 2, \ldots$

$$u(k) = \frac{C^{k}}{k!} \prod_{i=1}^{k-1}(i + \alpha) \tag{20}$$

The generating function for $u(k)$

$$U(t) = \sum_{k=0}^{\infty} u(k)\, t^{k} = \frac{1}{(1 - C\, t)^{\alpha}} \tag{18}$$

Distance function
with $u(1, k) = u(k)$ and $u(r, 0) = u(0) = 1$

$$u(r, j) = \frac{C^{j}}{j!} \prod_{v=1}^{j}(\alpha\, r + v - 1) \tag{86}$$

Probability $v(k)$, v_{k} of k error-free symbols after a symbol error

$$v(k) = u(k) \left(1 - \frac{k + \alpha}{k + 1} C\right) \tag{26}$$

Generating function for $v(k)$

$$V(t) = \frac{1}{t} + \frac{t - 1}{t\, (1 - C\, t)^{\alpha}} \tag{31a}$$

Correlation probability

$$r(s) = p_\text{e} \left(1 - \frac{\alpha}{1!} C - \frac{\alpha(1-\alpha)}{2!} C^2 - \cdots - \frac{\alpha(1-\alpha)\ldots(s-1-\alpha)}{s!} C^s \right) \qquad (35)$$

Generating function for $r(s)$
$$R(t) = p_\text{e} \frac{(1 - C\,t)^\alpha}{1 - t} \qquad (32)$$

4.3. Burst relationsships

Medium number of errors in a burst
$$E\{g\} = \frac{z_\text{f}}{z_\text{b}} = \frac{1}{u(a)} \qquad (38)$$

Probability of g errors in a burst
$$p\{g\} = u(a)\left[1 - u(a)\right]^{g-1} \qquad (39)$$

4.4. Disturbed block relationsships

Probability of single errors occuring in blocks

$$p\{n, g = 1\} = p_\text{e} \frac{2\alpha(2\alpha+1)\cdots(2\alpha+n-2)}{(n-1)!} C^{n-1} \quad \text{where } p\{n = 1, g = 1\} = p_\text{e} \qquad (68)$$

Probability of error bundle of length $l \geq 3$ in a block of length n

$$p\{l, n\} = p_\text{e}\, A\, B \qquad (73)$$
$$A = \frac{2\alpha(\alpha+1)\cdots(2\alpha+n-l-1)}{(n-l)!} C^{n-1}$$
$$B = 1 - \frac{\alpha}{1!} e^{-\beta} - \frac{\alpha(1-\alpha)}{2!} C^2 - \cdots - \frac{\alpha(1-\alpha)\ldots(l-2-\alpha)}{(l-1)!} C^{l-1}$$

Probability of errors in a block of length l and g errors in a block with $u(r,j)$ from (86) for $g = 2, 3, \ldots$ and $2 \leq g \leq l \leq n$

$$p\{l, g, n\} = p_\text{e}\, u(2, n-l) \sum_{r=1}^{g-1} \sum_{j=l-1-r}^{l-1} (-1)^{j+r+1-l} \binom{g-1}{r} \binom{r}{l-1-j} u(r,j) \qquad (92)$$

Probability of g errors in a block of length n termed "weight spectrum"

$$p\{g, n\} = \sum_{l=g}^{n} p\{l, g, n\} \qquad (95)$$

Bibliography

[1] Günther Söder. Digitalsignalübertragung: Digitale Kanalmodelle, Oktober 2016. URL `www.lntwww.de/downloads/6_Digitalsignaluebertragung/Theorie/Dig_Kap5_gesamt.pdf`.

[2] Claus Wilhelm. Der symmetrische Binärkanal mit Gedächtnis: Eine Systematik der Kanalmodelle. *Nachrichtentechnik-Elektronik*, 19(II.10):374–384, 1969.

[3] Claus Wilhelm. *Datenübertragung*. Militärverlag, Berlin, 1976.

[4] E. Johlig and u.a. Fehlerstrukturmessungen bei der Datenübertragung mit höheren Geschwindigkeiten. *Nachrichtentechnik-Elektronik*, 19(II.3):115–117, 1969.

[5] G. Hoffmann and u.a. Messtechnische Probleme bei Fehlerstrukturuntersuchungen im PCM-Kanal., 17. - 19.04.1973.

[6] Eugen Paul Rudolf Jahnke and Fritz Emde. *Tafeln höherer Funktionen*. Nauka, Moskau, 1968.

[7] D. Lochmann. *Statistische Untersuchung der Kombination von Stördetektor und Codierung bei der Datenübertragung im Fernsprechnetz am Beispiel von 200 bit/s*. Dissertation, Technische Universität, Dresden, 1971.

[8] Edgar Nelson Gilbert. Capacity of a Burst-Noise Channel. *Bell System Technical Journal*, 39(5):1253–1265, 1960. ISSN 00058580. doi: 10.1002/j.1538-7305.1960.tb03959.x. URL `https://archive.org/details/bstj39-5-1253`.

[9] Claus Wilhelm. Über den Zusammenhang zwischen Kanal und Codierung bei der Datenfernübertragung. *Nachrichtentechnik-Elektronik*, 17(10):368–392, 1967.

[10] William Wesley Peterson. *Prüfbare und korrigierbare Codes*. Oldenbourg, München, 1967.

[11] Siegfried Koller. *Graphische Tafeln zur Beurteilung statistischer Zahlen*. Theodor Steinkopff, Dresden and Leipzig, 2 edition, 1943.

[12] Andreas Ahrens and Claus Wilhelm. Bestimmung des Reduktionsfaktors bei Einsatz zyklischer Codes auf Kanälen mit gebündelt auftretenden Übertragungsfehlern. In Universität Rostock, editor, *10. Symposium Maritime Elektronik*, pages 141–144. Rostock, Juni 2001.

[13] John G. Proakis. *Digital Communications*. McGraw-Hill series in electrical and computer engineering. McGraw-Hill, Boston, 4 edition, 2001. ISBN 0072321113.

[14] William Turin. *Performance Analysis and Modeling of Digital Transmission Systems*. Kluwer Academic/Plenum Publishers, New York, 2004. ISBN 978-1-4613-4781-1. doi: 10.1007/978-1-4419-9070-9.

[15] Claus Wilhelm. *Konstruktion endlicher Kanalmodelle des symmetrischen Binärkanals mit Gedächtnis bei vorgegebenen nachrichtentechnischen Forderungen*. Dissertation, Technische Universität, Dresden, 1973.

[16] Claus Wilhelm. *Beiträge zur Bemessung von Nachrichtensystemen und zur Berechnung der Störfestigkeit*. Dissertation, Hochschule für Verkehrswesen, Dresden, 1989.